Finding Ghosts in Your Data

Anomaly Detection Techniques with Examples in Python

Kevin Feasel

Apress®

Finding Ghosts in Your Data: Anomaly Detection Techniques with Examples in Python

Kevin Feasel
DURHAM, NC, USA

ISBN-13 (pbk): 978-1-4842-8869-6
https://doi.org/10.1007/978-1-4842-8870-2

ISBN-13 (electronic): 978-1-4842-8870-2

Managing Director, Apress Media LLC: Welmoed Spahr
Acquisitions Editor: Jonathan Gennick
Development Editor: Laura Berendson
Coordinating Editor: Jill Balzano

Cover photo by Pawel Czerwinski on Unsplash

Distributed to the book trade worldwide by Springer Science+Business Media LLC, 1 New York Plaza, Suite 4600, New York, NY 10004. Phone 1-800-SPRINGER, fax (201) 348-4505, e-mail orders-ny@springer-sbm. com, or visit www.springeronline.com. Apress Media, LLC is a California LLC and the sole member (owner) is Springer Science + Business Media Finance Inc (SSBM Finance Inc). SSBM Finance Inc is a **Delaware** corporation.

For information on translations, please e-mail booktranslations@springernature.com; for reprint, paperback, or audio rights, please e-mail bookpermissions@springernature.com.

Apress titles may be purchased in bulk for academic, corporate, or promotional use. eBook versions and licenses are also available for most titles. For more information, reference our Print and eBook Bulk Sales web page at http://www.apress.com/bulk-sales.

Any source code or other supplementary material referenced by the author in this book is available to readers on GitHub. For more detailed information, please visit http://www.apress.com/source-code.

Printed on acid-free paper

To Mom and Dad, who know a thing or four about anomalies.

Table of Contents

About the Author

Kevin Feasel is a Microsoft Data Platform MVP and CTO at Faregame Inc., where he specializes in data analytics with T-SQL and R, forcing Spark clusters to do his bidding, fighting with Kafka, and pulling rabbits out of hats on demand. He is the lead contributor to Curated SQL, president of the Triangle Area SQL Server Users Group, and author of *PolyBase Revealed.* A resident of Durham, North Carolina, he can be found cycling the trails along the triangle whenever the weather's nice enough.

About the Technical Reviewer

Yin-Ting (Ting) Chou is currently a Data Engineer/Full-Stack Data Scientist at ChannelAdvisor. She has been a key member on several large-scale data science projects, including demand forecasting, anomaly detection, and social network analysis. Even though she is keen on data analysis, which drove her to obtain her master's degree in statistics from the University of Minnesota, Twin Cities, she also believes that the other key to success in a machine learning project is to have an efficient and effective system to support the whole model productizing process. To create the system, she is currently diving into the fields of MLOps and containers. For more information about her, visit www.yintingchou.com.

Introduction

Welcome to this book on anomaly detection! Over the course of this book, we are going to build an anomaly detection engine in Python. In order to do that, we must first answer the question, "What is an anomaly?" Such a question has a simple answer, but in providing the simple answer, we open the door to more questions, whose answers open yet more doors. This is the joy and curse of the academic world: we can always go a little bit further down the rabbit hole.

Before we start diving into rabbit holes, however, let's level-set expectations. All of the code in this book will be in Python. This is certainly not the only language you can use for the purpose—my esteemed technical reviewer, another colleague, and I wrote an anomaly detection engine using a combination of C# and R, so nothing requires that we use Python. We do cover language and other design choices in the book, so I'll spare you the rest here. As far as your comfort level with Python goes, the purpose of this book is not to teach you the language, so I will assume some familiarity with the language. I do, of course, provide context to the code we will write and will spend extra time on concepts that are less intuitive. Furthermore, all of the code we will use in the book is available in an accompanying GitHub repository at `https://github.com/Apress/finding-ghosts-in-your-data`.

My goal in this book is not just to write an anomaly detection engine—it is to straddle the line between the academic and development worlds. There is a rich literature around anomaly detection, but much of the literature is dense and steeped with formal logic. I want to bring you some of the best insights from that academic literature but expose it in a way that makes sense for the large majority of developers. For this reason, each part in the book will have at least one chapter dedicated to theory. In addition, most of the code-writing chapters also start with the theory because it isn't enough simply to type out a few commands or check a project's readme for a sample method call; I want to help you understand why something is important, when an approach can work, and when the approach may fail. Furthermore, should you wish to take your own dive into the literature, the bibliography at the end of the book includes a variety of academic resources.

Before I sign off and we jump into the book, I want to give a special thank you to my colleague and technical editor, Ting Chou. I have the utmost respect for Ting's skills, so much so that I tried to get her to coauthor the book with me! She did a lot to keep me on the right path and heavily influenced the final shape of this book, including certain choices of algorithms and parts of the tech stack that we will use. That said, any errors are, of course, mine and mine alone. Unfortunately.

If you have thoughts on the book or on anomaly detection, I'd love to hear from you. The easiest way to reach out is via email: `feasel@catallaxyservices.com`. In the meantime, I hope you enjoy the book.

PART I

What Is an Anomaly?

The goal of this first part of the book is to answer the question, "What is an anomaly?" Throughout this part of the book, we will learn what an anomaly is, some of the common use cases for anomaly detection, and just how powerful is that anomaly detection engine between your ears. We will cover a variety of psychological, technical, and statistical concepts, and by the end of this first part, we will have enough knowledge of anomalies and anomaly detection to begin writing code.

In Chapter 1, we will learn about outliers and anomalies as well as use cases for anomaly detection and one way to classify different types of anomaly detection techniques. Then, in Chapter 2, we will dive into psychology, specifically the Gestalt school of psychology. We will review some of the research around how the human brain reacts to patterns, as well as good ways to share information with people. Chapter 3 moves from the human aspect to computers. Humans are innately great pattern matchers, but computers are not. What we can do is train machines to understand what is normal and what deviates from normal. This third chapter provides us some of the statistical techniques we can use to train a computer to detect outliers.

PART I

What Is an Anomaly?

## CHAPTER 1

The Importance of Anomalies and Anomaly Detection

Before we set off on building an anomaly detector, it is important to understand what, specifically, an anomaly is. The first part of this chapter provides us with a basic understanding of anomalies. Then, in the second part of the chapter, we will look at use cases for anomaly detection across a variety of industries. In the third part of the chapter, we enumerate the three classes of anomaly detection that we will return to throughout the book, gaining a high-level understanding of each in preparation for deeper dives later. Finally, we will end this chapter with a few notes on what we should consider when building an anomaly detector of our own.

Defining Anomalies

Before we can begin building an anomaly detector, we need to understand what an anomaly is. Similarly, before we can define the concept of an anomaly, we need to understand outliers.

Outlier

An *outlier* is a data point that is significantly different from normal values. For example, during the 2008 Summer Olympics, Usain Bolt set a world record in the Men's 200 meters with a time of 19.30 seconds, finishing more than six tenths of a second faster than second-place Shawn Crawford. The remaining five competitors (excepting two runners who were disqualified) had times between 19.96 seconds and 20.59 seconds,

© Kevin Feasel 2022
K. Feasel, *Finding Ghosts in Your Data*, https://doi.org/10.1007/978-1-4842-8870-2_1

meaning that the difference between the second and the last place was slightly larger than the difference between the first and the second place. The results for the second through the sixth place are consequently *inliers*, or values that do not significantly differ from the norm. Table 1-1 shows the times of athletes in the final.

Table 1-1. *Completion times for the Men's 200 meters competition during the 2008 Summer Olympics (Source: `https://w.wiki/4CpZ` (Wikipedia))*

Athlete	Time (s)
Usain Bolt	19.30
Shawn Crawford	19.96
Walter Dix	19.98
Brian Dzingai	20.22
Christian Malcolm	20.40
Kim Collins	20.59

Based on this information, it is clear that Bolt's time was an outlier. The next question is, was it an anomaly?

Noise vs. Anomalies

Within the category of outliers, we define two subcategories: *noise* and *anomalies*. The difference between the two terms is ultimately a matter of human interest. Noise is an outlier that is not of interest to us, whereas anomalies are outliers that are interesting. This definition of anomalies matches the definition in Aggarwal (1–4) but is by no means universal—for example, Mehrotra et al. use the terms "outlier" and "anomaly" synonymously (Mehrotra et al., 4).

Another way of separating noise from anomalies is to think in terms of models with error terms. Suppose we have a function that takes a set of inputs and returns an estimation of how many seconds it will take that person to run the 200 meters competition. If we enter a variety of inputs, we can potentially map out the shape of this function, understanding that there is some variance between the estimation we generate and the true time. We show this by adding in an error term:

$$T_{Athlete} = f(A, B, \ldots) + \varepsilon$$

Occasionally, we might see a quarter-second difference from our expected time given some set of input values and the actual time. If we believe that this particular run still follows the same underlying function and the entirety of the difference is due to the error term, then the difference between our estimate and the actual time is, properly speaking, noise.

By contrast, suppose there is a separate function for a legendary athlete like Bolt:

$$T_{Bolt} = f(A, B, \ldots) + \varepsilon$$

This function provides considerably lower estimates of time than the run-of-the-mill Olympic athlete (if such a thing is possible!). If we use this function and get a result that is a quarter of a second faster than data points using the normal model, the outcome makes a lot of sense. If we were expecting the normal athlete output and ended up with the Bolt output, however, the result would be an anomaly: a result that is different from the norm and of interest to us because of the difference in underlying function used to generate the result.

In practice, this understanding of an anomaly is more probabilistic than concrete, as we do not necessarily know the particular function that generated any given data points, and this makes it difficult to separate noise from anomalies. In the Bolt case, we can expect that this is an anomaly rather than noise, not only because Bolt set a new world record for the event but also because his time was so much lower than the times of other world-class athletes, giving credence to the idea that he was a class apart.

Diagnosing an Example

Having seen one example already, let's review a graphical example. Figure 1-1 shows a scatter plot of points scored vs. yards passing in one week of National Football League games during the 2020 season.

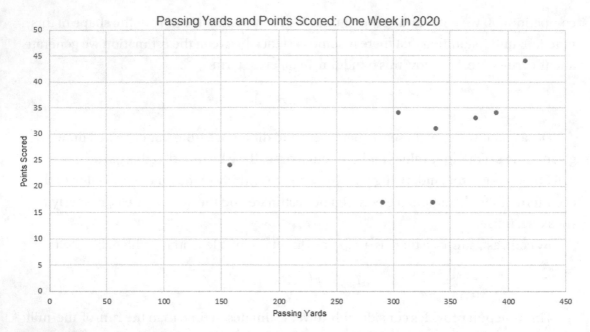

Figure 1-1. *Points scored vs. yards passing by the NFL team during one week in 2020*

As humans, we have certain innate and trained capabilities for analyzing this information and drawing conclusions from it, which we will cover in much greater detail in Chapter 2. For now, suffice it to say that there appears to be a positive correlation between number of passing yards and points scored in this plot.

Given the information available, we could imagine a fairly tight band that estimates the number of points scored for a given number of yards passing, drawing a thick line that comes close to or outright covers the majority of data points. Doing so would lead us to estimate that a team whose quarterback passes for 300 yards might score approximately 28–35 points on a normal day.

Having posited this theoretical model, we can then turn to the two outliers, in which the teams scored only 17 points despite their quarterbacks throwing for approximately 290 and 335 yards, respectively. This leads to two critical questions we need to answer—at least provisionally—in order to move forward with the process of anomaly detection. The first question is, is our model good enough to separate inliers from outliers properly? If the answer to this is "yes," then the second question is, how likely is it that the outlier data points follow a completely different process from the inliers?

Over the course of the book, we will look at ideas and techniques that will help us answer these two questions. That said, the second question may be quite difficult conclusively to answer, as it hinges on our understanding of the inner workings of the process that generates these data points.

This leads to another important consideration: domain knowledge is not necessary for *outlier detection*, but it is for *anomaly detection*. Drawing back to the example in Figure 1-1, a person with no knowledge of American football can perceive the relationship in the data as well as the two outliers in the dataset. What that person cannot do, however, is to draw a conclusion as to whether these data points are merely noise or if they imply a separate pattern of behavior. Making this concrete, if those teams only scored 17 points due to bad luck, missed field goals, or facing superior red zone defenses, these factors would point toward the outliers being noise. If, on the other hand, these teams have inefficient coaching strategies that move the ball down the field but consistently fail to convert that offensive momentum into points, then we may properly call these points anomalies. A domain expert with sufficient information could help us separate the two outcomes. As we will see later in this chapter, our intent in building an anomaly detector is to provide enough information for such a domain expert to analyze, interpret, and then act upon the signals we relay.

What If We're Wrong?

When deciding whether a point is an anomaly, we can perform the act of *classification* by splitting results into two or more classes based on some underlying characteristics. In the case of anomaly detection, we typically end up with two classes: this thing is an anomaly, or this thing is not an anomaly. If we have two possible classes (this is an outlier or this is not an outlier) and we can predict either of the two classes, we end up with a result like Table 1-2, which gives us the possible outcomes given a predicted class and its actual class.

Table 1-2. *Outcomes based on outlier predictions vs. a data point's actual status as an anomaly*

	Predicted Outlier	**Predicted Non-outlier**
Actual anomaly	Correct(true positive)	Type II error(false negative)
Actual Non-anomaly	Type I error(false positive)	Correct(true negative)

Looking at this table, we see two dimensions: whether a given data point was actually an anomaly or not and whether the prediction we made on that data point was positive (that yes, this is an outlier) or negative (no, this is not an outlier). When we correctly determine that an outlier is in fact an anomaly or a non-outlier is in fact a non-anomaly, we note that the classification was correct and can move on. There are two separate ways, however, in which we could go wrong. The first way we could go wrong is if we predicted that a data point would be an outlier, but it turned out not to be an anomaly—it was just noise. This is also known as a *Type I error* or a *false positive*. The converse of this is if we failed to state that some data point was an outlier despite its being an actual anomaly. The latter failure is known as a *Type II error* or a *false negative*. To the greatest extent possible, we want to avoid either type of error, but an interesting question arises from the fact that we classify these errors separately: Which of the two is worse?

The answer is the consultant's famous standby: it depends. In this case, the key consideration involves the negative ramifications of missing at least one actual anomaly and not taking action on it vs. receiving alerts or performing research on data points that turn out to be completely normal but were misclassified by our detection service. How many false positives you are willing to accept to limit the likelihood of a false negative will depend on a few factors, such as the frequency of false-positive activity, the level of effort needed to determine if a predicted outlier is actually an anomaly or not, and the cost of a false-negative result. Factors such as these will help you make the decision on how strict your outlier detector should be, though we will see an alternative approach starting in Chapter 4 that can help with this problem.

Anomalies in the Wild

In this section, we will briefly cover several industries in which anomaly detection plays an important role in decision-making. A detailed analysis of any of these industries is well beyond the scope of this work, but even a brief summary should suffice in laying out the thesis that anomaly detection is a cross-industry concern.

Finance

Anomaly detection serves several purposes in the financial sector. One of the most obvious scenarios for this pertains to pricing assets. The poster child for anomalous asset pricing has to be GameStop Corporation (GME), whose daily closing stock prices make up Figure 1-2.

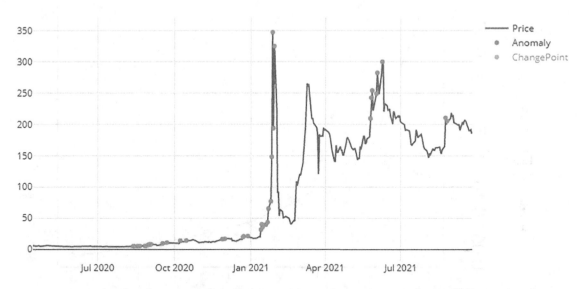

Figure 1-2. *Daily closing prices for GameStop Corporation (GME) from April 13, 2020, through September 24, 2021. Anomaly determinations come from the ML.NET library*

Relatively few stocks experience the turmoil that we saw with GameStop over a multimonth period, but based on information we have available at the time of writing this work, many of the outlier data points tie back to media campaigns (whether they be mass or social), due dates for options, and market movers taking sides on one of the largest market events of this generation.

Asset pricing is not the only place where anomaly detection makes sense in the financial industry. Another situation in which anomaly detection can be important is forensic accounting, that is, the review of accounting practices and financial documentation to determine if members at an organization have performed fraudulent actions. Forensic accounts do a lot more than pore over balance sheets and journal entries, but when they do look into financial documents, they are looking for patterns and anomalies.

One such pattern is Benford's Law, which states that the lead digit of a sufficiently large collection of numbers often follows a particular pattern, in which smaller numbers appear more frequently than larger numbers. Figure 1-3 provides that distribution in graphical form.

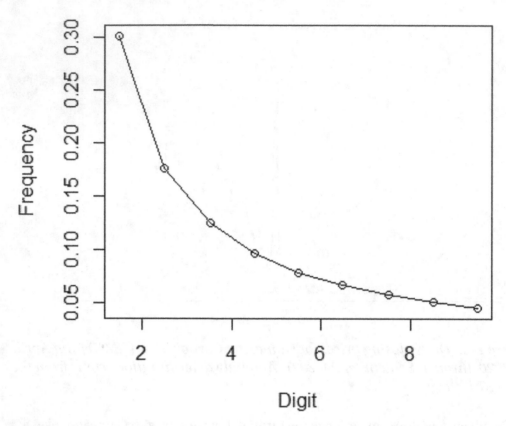

Figure 1-3. *A visual depiction of the distribution of the first digit of a sequence of numbers if it follows Benford's Law*

Not all datasets will follow this pattern, but a quick "rule-of-thumb" approach is that if a dataset includes data points spread across at least three (or preferably more) orders of magnitude, then it is likely to follow Benford's Law. This rule of thumb holds because many of the types of distributions that do adhere to Benford's Law—such as exponential or log-normal distributions—typically include a variety of values spread across several orders of magnitude, whereas distributions that do not adhere to Benford's Law— uniform and Gaussian distributions being the most common—often do not. For a detailed dive into the math behind Benford's Law, I recommend Hall (2019).

For our purposes, however, it is enough to understand that Benford's Law can provide a forensic accountant with expectations of how the distributions of certain datasets should look, particularly datasets such as the collection of journal entries for a corporation or governmental bureau. Considerable deviation from Benford's Law is not in itself proof of malfeasance, but it is a red flag for an auditor and typically merits further investigation to understand why this particular dataset is not typical.

Medicine

Anomaly detection is critical to the field of medicine. There exists a tremendous body of research on what constitutes healthy vs. unhealthy body measures. Examples of these measures include blood pressure, body temperature, blood oxygen saturation, and electrocardiogram (ECG) data. We can adduce that a person is currently in an unhealthy state based on significant changes in that person's measurements vs. the norm. Note that this norm could come as a result of aggregating and averaging data from a large population of individuals or it could come from cases in which the individual is in a state of known good health.

With modern electronics, we can collect and compare data in ways previously impossible outside of a hospital room. One example of a consumer product filling this niche is the Apple Watch, which is capable of tracking heart rate data and displaying it via an ECG app. This device can notify on abnormally high or low heart rates, as well as irregular heart rhythm. Mehrotra et al. provide other examples of where anomaly detection can be valuable in the field of medicine (11–12).

Sports Analytics

Sports is a big business around the world, with Forbes rating 12 sports franchises as worth at least $3.5 billion as of 2020. Further, depending on the sport, a team may invest $200 million or more per season in player salaries. In order to maximize this return on investment, teams have spent considerable time and money in the field of sports analytics. With this emphasis on analytics, it is important to analyze and understand outliers in order not to be fooled into thinking that statistical success in one season will necessarily carry over into the next.

A $23 Million Mistake

Chris Johnson was the starting third baseman for the Atlanta Braves coming into the 2013 season. He finished that season with a batting average of .321, which earned him a $23 million contract over the next four years. Coming into the 2014 season, the Braves expected him to be a key contributor on offense; instead, his batting average dropped to .263, and by late 2015, Atlanta general manager John Hart traded Chris Johnson as part of an exchange of bad contracts with the Cleveland Indians. At the time of his contract, there was good reason to believe that Johnson would not live up to it, and Figure 1-4 shows why, using a statistic called Batting Average on Balls in Play (BABIP). BABIP looks at how well a player does when putting the ball into play, that is, ignoring home runs as well as scenarios such as strikeouts or walks where the ball is never in play. A normal value for BABIP is typically in the .270–.330 range, but Chris Johnson's BABIP of .394 made him number 1 out of 383 qualified players with at least 150 at bats during the 2013 season.

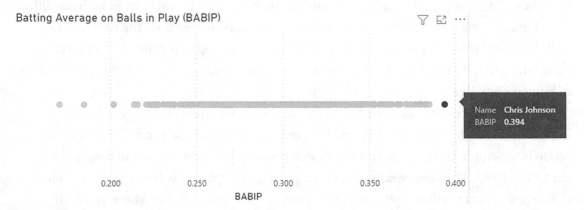

Figure 1-4. *Chris Johnson led Major League Baseball in 2013 with a BABIP of .394*

That Johnson led the league and had a BABIP of at least 60 points (i.e., 0.060) above the norm is important because there is considerable evidence that defenses are primarily responsible for determining whether a ball in play becomes an out, not the batter himself. There are exceptions to this general principle, but the implication is that typically, a player who has a very high BABIP in one season is unlikely to replicate this in the next season. In Johnson's case, he dropped from .394 in 2013 to .345 in 2014. Lacking considerable talents at hitting for power or getting on base, Johnson was more dependent on batting average for his offensive output than most players, and this reversion to a value closer to the mean was responsible for his offensive collapse. Had the Braves understood just how far of an outlier Johnson was in this important metric, they might not have offered him a bad contract after his 2013 season, letting some other team make that mistake instead.

A Persistent Anomaly

Greg Maddux is one of the greatest pitchers of all time. He continually stymied batters for 23 seasons in the major leagues, thanks to the incredible control of his pitches and an almost-unparalleled ability to change speeds on pitches without tipping it off to batters.

Maddux also holds the honor of being the pitcher with the greatest differential between his Earned Run Average (ERA) and his Fielding Independent Pitching (FIP) ratings for players between 1988 and 2007, a range that covers the point in which Maddux turned the corner as a young starter to the year before he hung up the cleats for good.

FIP as a measurement originated in research on understanding just what factors a pitcher is able to control vs. his defense. The result of this research is that pitchers are primarily responsible for three things: walks, strikeouts, and home runs. Otherwise, the majority of contribution to batted balls in play lands squarely on the defense—just like what we saw with BABIP in the prior section.

One conclusion of FIP is that it tends to be more predictive of a pitcher's ERA in the next season than the pitcher's ERA in the current season, as defenses tend toward the mean. This means that if you see a pitcher whose ERA is considerably lower than his FIP, we can expect regression the next year. That is, unless he is Greg Maddux. Figure 1-5 lays out 1703 seasons from pitchers and shows the difference between their ERA and FIP, where a negative value means that a pitcher outperformed his FIP and a positive value means he underperformed.

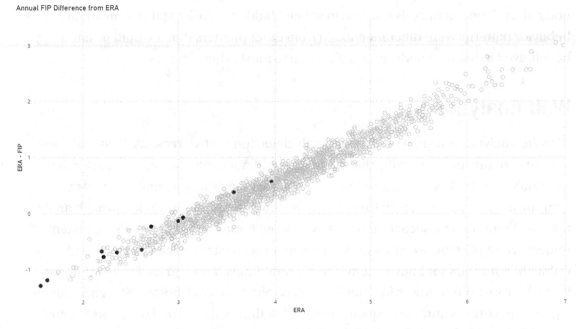

Figure 1-5. *ERA is plotted on the X axis and the difference between ERA and FIP on the Y. Greg Maddux's seasons with the Atlanta Braves, during the peak of his career, are highlighted in black*

Greg Maddux is responsible for the two biggest differences between ERA and FIP—and they came in consecutive years to boot, meaning that he did not regress to the mean as you would expect. Maddux is responsible for four of the top 30 player seasons in this time frame and consistently outperformed his FIP. Figure 1-6 shows the sum of differences between ERA and FIP over time for all pitchers between 1988 and 2007, and Maddux is at the top of the list, followed by Pedro Martinez, Jose Rijo, and Dennis Martinez.

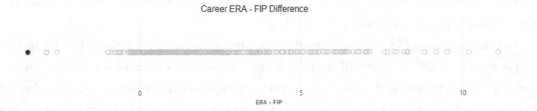

Figure 1-6. *Career summations of ERA-FIP for all pitchers between 1998 and 2007*

In this case, we once again see an outlier: Maddux consistently outperformed on his ERA compared to FIP expectations, something very few pitchers can claim. Based on the body of work he put up, this is a case in which Maddux's outlier status is an anomaly: his behavior patterns were different enough from other pitchers that it would be fair to say he followed a different performance function to most other pitchers.

Web Analytics

Website analytics is a great field for anomaly detection for two reasons. First, websites tend to generate enough traffic that a human could not parse through the data in any reasonable time. This necessitates the use of log aggregation and analytics tools to simplify the job, but even with these tools, the time requirements for a thorough analysis may be daunting. The second reason is even more important: there exists a constant undercurrent of errors on any site. Automated tools constantly scan websites and public IP addresses for known, common vulnerabilities. This can lead to 404 (file not found) errors from requests for a file like `index.php` on a .NET-hosted site. These are not legitimate requests and are frequent enough that they will overload with noise in any naïve system that broadcasts alerts assuming you do not filter these messages out.

It may be enough to filter out illegitimate errors, but there exist at least two more classes of uninteresting error: outdated links and accidental errors. For an example of the former, suppose I send an email with a link to a photograph I have put online.

Some number of years later, I have deleted the photograph from the website, but you still have the link in your email. If you open that link, you will get a 404 (file not found) error, as the resource no longer exists. For an example of the latter, I may attempt to access a URL from memory and, during the process, accidentally strike the wrong key, leading to a typo. This could lead to a 404 error as well. Similarly, I might accidentally click on a protected link, be prompted for credentials, and then select Cancel, leading to a 403 (not authorized) error. These are legitimate errors, but from the standpoint of a systems administrator, this is normal and expected behavior. There is nothing for the systems administrator to "fix" because everything is working as designed.

Where anomaly detection can be interesting is in finding errors above and beyond the "background radiation of the Internet" (with credit to Steve Gibson for the analogy) and common miscues. Suppose a new product release goes out and the development team has enabled a feature flag that sends 1% of traffic over to a new page, leaving 99% to stay on the existing page. Suppose further that there is a problem on the new page, such that some percentage of people receive 404 errors pertaining to missing resources on the page. Now we have a different pattern of behavior—it is no longer one or two failed requests for a given resource, nor is it broad-based noise. An anomaly detection system that can find these sorts of errors and alert on them while filtering out most of the preceding noise can save systems administrators and development teams considerable time while still finding code release issues before a customer issues a trouble ticket.

And Many More

By this point, it should be clear that anomaly detection is not limited to a single industry or field of study but instead is broadly applicable. Furthermore, it turns out that there are a common set of anomaly detection techniques that are independent of field or industry, meaning that the techniques we learn in this book will be applicable regardless of your industry.

Now that we have completed a brief survey of anomalies in the wild, the next step is to look at the three classes of anomaly detection.

Classes of Anomaly Detection

There are three broad classes of anomaly detection technique available to us. We will look at each in turn, based on the order in which they appear in subsequent chapters. All three techniques are valuable in their own domains, making it worthwhile to understand the key components of each in order to apply the best approach to a given problem.

Statistical Anomaly Detection

The first class of techniques for anomaly detection is statistical anomaly detection techniques. With these techniques, we assume that the data follows some particular underlying distribution—for now, think of this as the shape of a dataset—but other than the distribution of the data, we make no assumptions about what this data "should" look like. In other words, we do not have a model in which, given some set of inputs, we will get some specific predicted output.

For statistical anomaly detection techniques, outliers are cases that are significantly distant from the center of our distribution—that is, they are far enough away from the center that we can expect that they aren't really part of the same distribution. The definitions of "significantly different" and "far enough away" are necessarily loose, as they may be user-configurable and likely depend on the domain.

Chapters 6 through 8 focus on incorporating statistical techniques into a generalized anomaly detection service.

Clustering Anomaly Detection

The second class of techniques for anomaly detection focuses on clustering. The concept of clustering is that we group together points based on their distance from one another. When points are sufficiently close to one another, we call those groupings *clusters*. We make no assumptions regarding underlying relationships in the data—that is, how individual data points end up clustered together. We do not assume that the data follows any particular distribution, nor do we attempt to build a model of inputs relating to outputs. Instead, we are concerned only with the outputs and the clusters they form.

When performing outlier detection using clustering techniques, we consider a data point to be an outlier if it is sufficiently far from any given cluster. This makes sense given our definition of an outlier as a point sufficiently distant from the norm but requires us to expand our understanding of "the norm" from a single point—like the center of a distribution—to a set of points, each of which may be independent of the others.

Chapters 9 through 11 will focus on incorporating clustering techniques into a generalized anomaly detection service, and Chapter 12 introduces an interesting variant on clustering.

Model-Based Anomaly Detection

The final class of techniques we will cover in this book is modeling techniques. In this case, we train some sort of model of expected data behavior. Typically, these models follow a similar path: given some input or set of inputs, we assign weights for each input based on how valuable it is and then combine those weighted inputs together in some way to generate our output. This technique is known as *regression* and is extremely common in the world of data science. Our formula for predicting a sprinter's time is an example of regression:

$$T_{Athlete} = f\left(A, B, \ldots\right) + \varepsilon$$

In this case, we do not specify exactly what the weights are or how we combine variables, as those are implementation details and will depend on the nature of the specific problem. As an example, we might assume that predicting an Olympic athlete's time is based off of a linear model, in which each weighted variable is summed together to come to an estimation of the time, like so:

$$T_{Athlete} = w_1 x_1 + w_2 x_2 + \ldots + w_n x_n + \varepsilon$$

Outliers in this case are cases in which the actual calculated data point is substantially different from the predicted data point. Going back to the case of Usain Bolt, we might plug in our sets of weighted inputs and (for example) get a time of 20.10 seconds for the Men's 200 meters competition. When he runs it in 19.30 seconds, our prediction is far enough off that Bolt's time becomes an outlier. Then, we can make the leap in intuition that Bolt does not follow our normal model but instead follows a different model, one that might have different weights on the input variables or may even include variables not relevant to others. Whatever the case may be, once we assume that Bolt follows a different model, we can now assert that those outliers are in fact anomalies: they were generated from a different process than the one that generates the majority of our data.

Making this leap requires one key assumption: the model we select fits the behavior of the underlying phenomenon sufficiently well in the large majority of cases. If our model is not good at predicting results, we will end up with incorrect depictions of outliers, as the model is insufficiently good at determining what constitutes normal behavior. If that is the case, we should not trust the model to tell us that something is an outlier.

Chapters 13 through 17 focus on one particularly interesting variant of model-based anomaly detection: time series analysis.

Building an Anomaly Detector

Now that we have a high-level idea concerning what an anomaly is and what techniques we might use to find anomalies, it is time to begin the process of building an anomaly detection service, something we will work toward throughout this book. Before we write a line of code or dive into research papers, however, we need to gain a better understanding of how our customers might make use of an anomaly detection service. For now, we will look at three key design concepts.

Key Goals

The first thing we will need to understand is, what is the purpose of our anomaly detection service? There are several reasons why a person might want to make use of an anomaly detection service. The first reason is to alert a human when there is a problem. For example, going back to the web analytics case, we want to alert systems administrators and developers when there is a problem causing people not to see relevant web resources. At the same time, we do not want to alert staff to things like automated bots poking for common vulnerabilities (and failing to find them) or people mistyping URLs.

A second reason, entirely separate from alerting, is to test a conjecture. Table 1-1 showed us just how fast Usain Bolt was during the 2008 Olympics, and we could see that the difference in times between Bolt and the rest of the competitors was drastic. That said, we may wish to see a more objective analysis of the data, something that might determine the probability that a normal Olympic athlete could run 200 meters in Bolt's 19.30 seconds. This backward-looking analysis does not necessarily lead to some subsequent alerting requirement, marking it as a separate use case.

A third reason why we might want to use an anomaly detection service does follow from the alerting scenario in the first use case, but it also goes one step further: we could use an anomaly detection service to allow a program to fix an issue without human intervention. For example, suppose you sell high-end computer monitors with an average retail value of $1,500 USD. A major holiday is coming up and you want to discount the cost of your monitor by $150 USD, selling it for a total of $1,350 USD. Unfortunately, in trying to apply the discount, you set the *price* to $150 USD, and a large number of orders come in as people find this monitor marked at 90% off its retail price. This leaves the company in a bad position, needing either to allow the orders and absorbing a major loss or cancelling the orders and potentially raising the ire of customers as well as the shopping platform—in the latter case, potentially jeopardizing the customer's relationship with the platform to the point where the customer gets banned from selling on that platform. An alerting service might indicate that the price range for this monitor has never strayed from $1300–1500 USD and this new price is radically different. We could imagine a system that prevents the price change from going out before a human confirms that this is, in fact, the intended price.

Finally, we may not even know what "normal" looks like and wish to discover this. For example, suppose we wish to find out what type of person is most likely to purchase a $1,500 monitor. Given demographic details, we might find that there are certain clusters of individuals who tend to purchase our monitors and we might market toward those customers. The outliers from this process will also be interesting because they might be indicative of potential groups of customers who have not purchased our monitor but might be inclined to if we understand more about their desires. Given information on outliers, our analysts may perform market research and determine if there are untapped segments of the market available and, if so, how to reach these people.

How Do Humans Handle Anomalies?

The next consideration is to ask yourself, how will the humans who use your anomaly detection service handle reports of outliers and determine whether a given outlier is actually an anomaly or if it is merely noise? For example, is there a specific investigative process that a human should follow? In a manufacturing scenario, we might imagine anomaly detection on sensors such as machine temperature or the amount of pressure a given machine exudes to press and fold a box. Suppose that a human observes the machine temperature and, if the temperature rises beyond an acceptable level, the human presses a button to dispense a cooling agent. If that action does not lower the

19

temperature to an acceptable level, the human then presses another button to slow down the operation. Finally, if the temperature still remains well outside of normal parameters, the human presses yet another button to halt the line until the machine has cooled down or a technician has completed an investigation of the machine.

The reason it is important to understand how people will use your anomaly detection service is that you might be able to help them automate certain processes. In the preceding scenario, we could easily imagine a system feeding temperature data to our anomaly detection service and, upon receipt of a certain number of outliers in a given time frame, automatically perform each of the three steps. Providing a more nuanced solution, perhaps the service would automate the dispensation of a cooling agent while writing a warning to a log. Then, if the system slows down the operation, it might alert the technicians on staff to make them aware of the problem, leaving the option of complete shutdown to a human.

In the event that we cannot automate this kind of activity, users may still find value in having an anomaly detection service because it could provide the basis for an investigation. Returning to the website analytics example, an abnormal number of errors—whether that be an abnormally large number of errors in general or an abnormal number of a particular class of error like 404 (file not found) or 403 (not authorized)—could give systems administrators enough information to dive into the logs and track down what might have caused a change in behavior, be it a code deployment, infrastructure change, or failing service.

In between these two levels of interactivity, we could also anticipate having a service that depends on the anomaly detection service and, upon receipt of an outlier, queues something for manual review. An example of this would be manual review of automated price changes on products. The large majority of the time, we might not want manual reviews of changes because the number of products changed can be quite large. An incident in which the price of a product is dramatically different from its norm, however, would be a great case for manual review before the actual price change occurs. This might save a company considerable pain down the line.

When trying to discern the specific intentions of the callers of your service, the most important question is, what domain knowledge is necessary in humans when you alert them to a potential anomaly? An anomaly is, using our accepted definition, necessarily interesting. What information does a human have that makes this thing interesting and not just another incident of noise? Furthermore, how much of that information can we automate or transfer to machines? Not everything is amenable to automation, whether due to a lack of appropriate equipment or because of the amount of disparate, subjective knowledge in the heads of a variety of people preventing an automated solution from being viable. That said, the typical reason for existence of an anomaly detection service is to provide information on potential anomalies for subsequent investigation and potential action.

Known Unknowns

As we wrap up this discussion of anomalies, we have one final consideration: How much information do we already have on existing anomalies? Suppose that we have a history of all temperature recordings for a given machine. Further, an expert has combed through each of those recordings and determined whether or not the machine behaved incorrectly, allowing us to correlate high temperature with improperly machined products that we needed to repair or scrap. Armed with this information, we can perform *supervised learning*, a subset of machine learning that uses known good information to draw conclusions on unknown information. Supervised learning is a particularly popular class of machine learning in the business world—the majority of machine learning problems tend to fit supervised learning scenarios, and there exists a wealth of tooling and research into supervised learning techniques.

Unfortunately for us, anomaly detection tends not to fit into the world of supervised learning, as we rarely have the luxury of experts building comprehensive datasets for us. Most often, we have to perform *unsupervised learning*, another class of machine learning in which we do not have known good answers as to whether a particular data point was an anomaly. Instead, it is up to us to discover whether that given data point was anomalous. As we will see throughout this book, there are plenty of techniques we can use for unsupervised learning, but all of them come with the caveat that we cannot easily determine how correct (or incorrect) our anomaly detector is, as we do not have good information on what constitutes an anomaly vs. a non-anomaly.

Sometimes, we may have knowledge of some prior anomalous scenarios. This is a middle ground between the two variants known as *semi-supervised learning*. The idea here is that we may have a tiny amount of known good data, such as a recorded history of some anomalies, and use this information to train a model as we do in the supervised scenario. From this tiny sample, we use clustering techniques to determine which data points are sufficiently similar to our known anomalies and proceed to label these similar data points as anomalies. We can show those points marked as outliers to humans and have them tell us whether our predictions were right or not, thereby expanding our known input data without using a lot of time and energy to review large amounts of data. Semi-supervised learning is typically more accurate than unsupervised learning, assuming you have enough valid data to create an initial model.

Conclusion

Over the course of this first chapter, we laid out the nature of outliers, noise, and anomalies. Outliers are values that differ sufficiently from the norm. Noise and anomalies are partitions of the set of outliers. Outliers that are not of interest are noise, whereas interesting outliers are anomalies. Anomaly detection is important wherever there is statistical analysis of problems, meaning that anomaly detection has a home in a wide array of industries and fields.

There are three classes of anomaly detection technique that we will focus on throughout the book: statistical, clustering, and model-based techniques. Each class of technique brings its own assumptions of the data and scenario, and each is valuable in its own regard. We will see that certain types of problems are more amenable to one class of technique than the others, so having a good understanding of all three classes is important.

Finally, we looked at the human element of anomaly detection. No matter how well your service detects anomalies, it is useful inasmuch as it provides value to humans. To that extent, understanding the needs of those who use your service can help you build something of great value. Although we will focus on the technical aspects of outlier detection and software development, be sure to keep your end user in mind when developing the real thing.

## CHAPTER 2

Humans Are Pattern Matchers

In the prior chapter, we learned about *outliers*, *noise*, and *anomalies*. We also made use of humans' intuitive ability to detect anomalies. In this chapter, we will learn more about how humans interpret and process information. In particular, we will focus on the findings of the Gestalt school of psychology.

A Primer on the Gestalt School

In the year 1912, Max Wertheimer wrote a paper entitled *Experimentelle Studien über das Sehen von Bewegung* (translated, "Experimental studies on the seeing of motion"). In this paper, Wertheimer describes the notion of *phi motion*, in which a person observes motion despite the object not moving. Wertheimer, along with colleagues Kurt Koffka and Wolfgang Köhler, subsequently founded the Berlin school of Gestalt psychology, which we will refer to as the Gestalt school. The term *Gestalt* translates to "form" or "figure" and represents one of the major underpinnings of their theory: that our minds operate in terms of structured wholes rather than individual sensations. Put overly simply, people observe things based on mental patterns. More specifically, people interpret things in the simplest possible way, meaning that we interpret arrangements to have fewer instead of more elements, symmetrical instead of asymmetrical compositions, and to look like things we have observed before. This is the Law of *Prägnanz* (translated, "conciseness" or "terseness"), best described in Koffka's *Principles of Gestalt Psychology*.

The Gestalt school has seen its shares of ups and downs in the intellectual community, with a series of critiques and revivals over the past century. This book will not adjudicate the claims and counterclaims of researchers in the field of psychology, but we will take the findings with the most experimental support behind them and use this as a guide for understanding human behavior.

© Kevin Feasel 2022
K. Feasel, *Finding Ghosts in Your Data*, https://doi.org/10.1007/978-1-4842-8870-2_2

Key Findings of the Gestalt School

The goal in this section is to review four key Gestalt school findings that form the concept of Prägnanz. Then, we will review 12 implications of these findings and see how humans use these principles to observe visual elements and draw conclusions based on these observations.

Emergence

The first key finding in the Gestalt school is the concept of *emergence*. Emergence is the notion that we perceive things immediately and without conscious effort when the brain recognizes a familiar pattern. For example, I may perceive that a particular cloud looks like a giraffe. This perception occurs without me explicitly thinking that I wish to see a giraffe but is instead automatic. A person who has never seen a giraffe (either in person or a pictorial representation of one) will not observe the cloud in the same way that I will.

Quadrupedal mammals are, of course, not the only things our brain will unconsciously perceive. Another common example we see is faces. Figure 2-1, for example, is a photograph of a rock formation taken in Tianjin, China. From the moment I saw this rock, I knew that it looked remarkably like the economist and moral philosopher Adam Smith.

Figure 2-1. *A rock formation in Tianjin, China, bearing a resemblance to Adam Smith*

Our brains are capable of rejecting associations later on, but that immediate recognition does occur. I might end up deciding that the rock in Figure 2-1 does not, in fact, look like Adam Smith (though it really does!), but that did not stop the immediate, pre-conscious interpretation from occurring.

Reification

The second key finding ties in nicely with the first, and that is the concept of *reification*. Reification, in this context, means that we perceive whole, meaningful, concrete objects upon recognition. Further, when we do recognize a particular object, it takes additional mental effort to focus on abstract shapes.

For example, humans are much better at picking a familiar person out from a crowd than raw probabilities might first assume. Finding a group of friends in a crowded restaurant is surprisingly easy, even when parts of bodies or faces are blocked from view by tables or passers-by. We are also great at noticing people or things even when they are partially obscured, such as a person mostly covered by a quilt or a kitten trying to hide by covering its eyes with its paws.

A joke built off of the concept of reification comes from a season 8 episode of *The Simpsons* entitled "In Marge We Trust," in which Homer sees his likeness in a Japanese detergent mascot, Mr. Sparkle. The mascot comes from the combination of Matsumura Fishworks (represented by a smiling fish) and Tamaribuchi Heavy Manufacturing Concern (a light bulb). Because audiences have seen the face of Homer Simpson for more than seven years, they will naturally associate the "fishbulb" with Homer.

Reification also explains the key behind camouflage. Camouflage is not intended to make you invisible, strictly speaking; instead, its purpose is to break up your natural shape and make it harder for a person to observe you in your entirety. In nature, camouflage appears in several guises, such as stripes (tigers), spots (leopards), and countershading (gazelle). For humans, camouflage typically comes in military or hunting contexts, and the specific patterns will depend upon the terrain. A particularly interesting historical example was "dazzle camouflage," a series of high-contrast, intersecting geometric shapes painted on ships during World War I. The intent of dazzle camouflage was to make it harder for German submarines to tell the speed and heading of a particular ship, thereby making it harder for German submarines to sink British and American vessels. In this case, the camouflage actually made it easier to see that a ship was there but (at least in theory—results were inconclusive as to whether dazzle camouflage actually worked) made it more difficult to target and sink the ship.

Invariance

The third finding of note is the concept of *invariance*. Invariance means that we will recognize most shapes and figures despite some levels of variation. For example, I know what a circle looks like and can recognize a circle even if this circle is larger than the prior one I saw or has a different fill color. I can observe a dog I've never seen before and know that this is a dog. Furthermore, I can recognize—sometimes with difficulty—the same thing or person at different points in time. If I have not seen an old classmate for a decade, I may still recognize the person despite a different hairstyle, manner of dress, and waistline situation. What is happening here is that our brains are able to extract the basic characteristics of a person or thing and allow us to generalize this to other incidents and times.

Invariance is not the same as perfect recall or understanding, however. Having seen the cover of a DVD a dozen years ago, I might not recognize it today. Typically, the people or things we see most consistently or which bond most closely to us will be recognizable over time, even with significant variation.

Multistability

Our final key finding is called *multistability*. The concept of multistability is that, if we lack sufficient visual cues to resolve a particular situation, we will alternate between different interpretations of a scene. For example, Figure 2-2 shows a classic example known in the literature as Rubin's vase. Rubin's vase is named after psychologist Edgar Rubin, himself not a Gestalt school psychologist but a colleague and inspiration to Koffka et al.

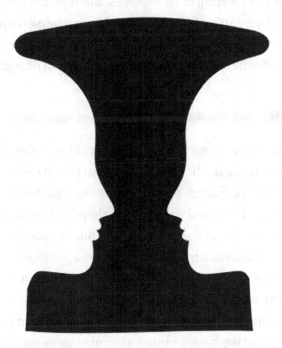

Figure 2-2. *A variant of Rubin's vase by Yamachem, hosted on Open Clipart*

When viewing this image, our brain can interpret two distinct images: this either is a picture of a single vase or it is two faces pointed at one another. Because both images are possible and make sense to our brains, we have trouble deciding which of them is the "correct" image and therefore each one becomes ascendant in our minds for some period of time, only to be superseded by the other. This continues until we have a sufficient visual cue to disambiguate the image. In the case of Rubin's vase, this is typically the addition of a third dimension, where we can clearly see either the depth of the vase or the depth of the faces. Adding this depth eliminates any ambiguity, making it much more difficult to see the "wrong" answer.

Principles Implied in the Key Findings

The four key findings are interesting in themselves, but they also lead to a series of follow-up principles that we can use to understand how a person interprets the signals that the eyes send to the brain. In total, there are 12 principles we will review in this section, laid out in an order that allows principles to build on one another.

Meaningfulness

The first principle we can derive from the preceding findings is the principle of *meaningfulness*. Meaningfulness is the idea that we can group visual elements together to form meaningful or relevant scenes. This principle comes straight out of the concept of emergence. In Figure 2-1, I grouped together the shapes of several rocks, interpreted a human face, and gave meaning to the formation. Similarly, people looking at clouds will often see shapes or animals as a by-product of this capability. In the context of data analysis, we can view a scatter plot and discern a trend in the data. The trend itself is not a visual element but is instead something we impute based on prior experience or intuition.

The flip side of meaningfulness is that we may ascribe certain properties to things that are not factual. Suppose we have a random number generator that generates integers between 1 and 100 with equal probability and it prints out the sequence {8, 9, 10}. We may see a trend and assume the next value will be 11, but supposing that the random number generator is not broken, we have a 99% chance of the next number *not* being 11. In this case, we have spotted a spurious trend—or, to use the language of Chapter 1, we have a sequence that forms an outlier, but it turns out to be noise rather than a proper anomaly.

Conciseness

The second principle is that of *conciseness*. Humans have a tendency to reduce reality to the simplest form possible. For example, Figure 2-3 shows two images. On the left-hand side, we see an image. The right-hand side shows the same image, but with each shape having a different color. In both cases, we immediately perceive that the "shape" on the left is really the combination of a triangle, a circle, and a square, as we see on the right.

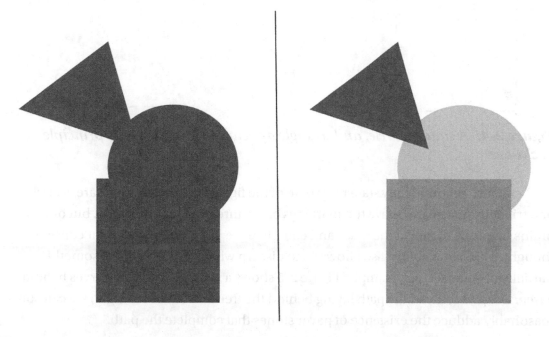

Figure 2-3. *Even when shapes overlap and lack defining color differences or borders, as on the left-hand side, we still find a way to break them down to the smallest meaningful element*

The reason we see three shapes rather than one is that we don't have a mental picture of something that looks like the thing on the left; therefore, we break it down into component parts that we do understand. Similarly, if I show you a picture of a cat lounging on the arm of an upholstered chair, you will see a chair and a cat, not some beast that is the agglomeration of the two. This principle ties back to the concept of reification, in which we perceive whole, meaningful, and concrete objects wherever we can. If we did frequently see the image on the left as a caricature of a person with a great mohawk but horribly receding hairline, we would begin to perceive it as a whole object, as we now can ascribe meaning to it independent of the underlying shapes.

Closure

The third principle is *closure,* in which the brain automatically supplements missing visual information in order to complete the picture. Figure 2-4 shows an example of closure, in which I can confidently state that you will see a square, a circle, and a triangle, respectively.

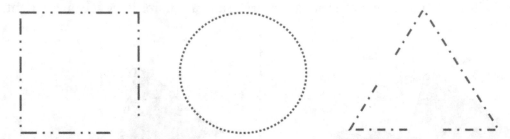

Figure 2-4. *A square, circle, and triangle appear in view due to the principle of closure*

I can just as confidently state that none of the figures are, in fact, a square, a circle, or a triangle. Instead, we have the markings of the three shapes broken up, but our brains helpfully "connect the dots" and complete the pattern for us without conscious thought. This same ability also allows us to discern what a thing is even if something partially obscures it. For example, Figure 2-5 shows a walking path that curves behind a tree. Despite some of the path laying behind the tree and out of our sight, we can still reasonably adduce the existence of paver stones that complete the path.

Figure 2-5. *Our brains "complete" the part of the path that winds behind the tree*

Similarity

One of the most powerful principles is that of *similarity*. This principle finds that we will naturally group items based on color, size, shape, and other visual characteristics. Furthermore, when it comes to color, we can include hue, saturation, brightness, vibrance, and temperature in our discussion of what defines similarity. When showing images in grayscale, we lose much of that information, but Figure 2-6 is nonetheless clear enough, allowing us to see three separate colors (or shades of gray in print).

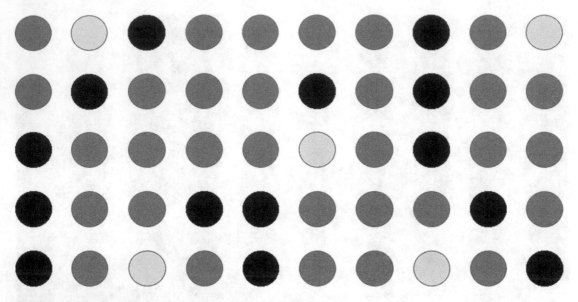

Figure 2-6. *The principle of similarity in action*

As we look at the image in Figure 2-6, people tend, absent any other information, to assume that dots of the same color have some common meaning, such that the black dots scattered across the image all represent something independent of the blue and yellow dots. From there, we could introduce additional complexities, such as changing some of the dots to squares and triangles. Then, we would assume that a similar color and shape represents a single thing and perhaps that shapes of the same color have some similarity but are not quite equivalent. Going even further, I might include larger and smaller shapes of the various colors, and again, absent any other information, we tend to assume that similarly sized elements share something in common with one another. As we add more layers of complexity, the assumptions pile up and you are more likely to confuse the viewer rather than elucidate the situation. Even so, with a few hints, our brains can process information across several visual dimensions with no problem.

Good Continuation

The fifth principle is *good continuation*, in which our brains want to follow smooth, predictable paths. Figure 2-7 shows an example of the principle of good continuation. Here, we use the property of closure to see two lines out of a series of circles. The topmost line curves downward and intersects with a straight line in the middle of the image.

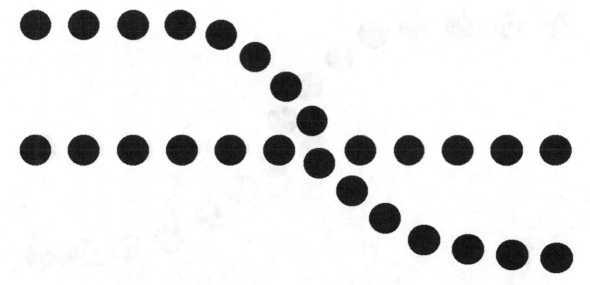

Figure 2-7. *The principles of closure and good continuation have us see one straight line and one curved line*

Despite there not being an actual line in the image, we can easily see the makings of the two lines from the spacing of the circles that do appear, and our brains can interpret the scene accordingly. Furthermore, we can assume that others will interpret the lines in the same way—very few people will see three (or more) lines or imagine that the two lines fail to intersect.

In order to ensure that others do see two lines failing to intersect, we can introduce color into the mix and take advantage of the principle of similarity. Now, in Figure 2-8, we have two sequences of dots: black and blue (or gray in print).

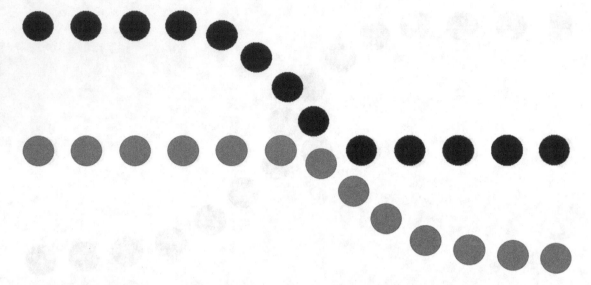

Figure 2-8. *Our brains naturally rebel against discontinuities in images*

With the change in colors, we reluctantly see lines curving away from one another. This also gives us an indicator that some principles are stronger than others: similarity, especially with respect to color, overwhelms most of the other principles and allows us to guide the viewer in spite of other inclinations. This is also a great summary of why you should not overuse color in visual design, as by doing so, you lose the ability to draw a viewer's attention and correct bad assumptions.

Figure and Ground

Another principle closely related to the Gestalt school is the duality of *figure* and *ground*. The figure is the subject of our image, and ground represents the setting around the subject. We can use visual cues to decide what is the figure of importance vs. what constitutes the ground. Generally, one of the best cues is that the figure tends to be smaller relative to the ground. Another cue is when figure and ground contrast sharply with one another with respect to color.

Tying back to the key findings of the Gestalt school, the concept of multistability is something we can better state as a confusion between the active figure and the ground. Rubin's vase, as in Figure 2-2, is an example in which we have two viable possibilities for the figure, leading us to see each one in turn for some amount of time. Interestingly, even in this case of multistability, we don't end up in a situation in which both the faces and the vase are the figure at the same time—one must be the ground, while the other becomes the figure.

Proximity

The principle of *proximity* indicates that we will perceive things grouped closely together to be more related than things that are spread further apart. For example, Figure 2-9 shows a series of dots spaced along a line. Each of the dots has exactly the same color and size, yet we see four separate groupings due entirely to the spacing between dots.

Figure 2-9. *Despite being equally sized dots of the same color, we perceive four groups based on proximity*

This principle of proximity will tie into Part III of the book, in which we cover distance-based anomaly detection techniques. Distance-based anomaly detection techniques rely on the assumption that data points that are closer together are similar and if we have a sufficient number of data points sufficiently close together, all of those points are inliers.

Connectedness

Connectedness is the idea that we perceive physically connected items as a group. As a simple example, imagine a classic yellow Number 2 pencil. We speak of a pencil as a thing in itself, and yet it is made up of a graphite core, a wooden frame surrounding the graphite, a rubber eraser, a metal band attaching the eraser to the frame, and a coating of yellow paint on the wood. We can, of course, focus in on any one of these components when necessary, but our default interpretation is the pencil as a whole. Connectedness acts as the culmination of the principle of proximity.

Common Region

The principle of *common region* takes us one step further than the principle of proximity. With common region, we assume that things that are in a clearly delineated region of space are necessarily similar. Figure 2-10 shows us an example of this phenomenon. In this case, we have a regular sequence of dots, all of which share the same color, size, and shape. They are also laid out equidistantly from one another, making this as homogenous a visual as possible. Then, we draw a box, which acts as a background behind a cluster of the dots. All of the dots inside the box now appear to be part of one common region, leaving the remaining dots in a second, box-free region.

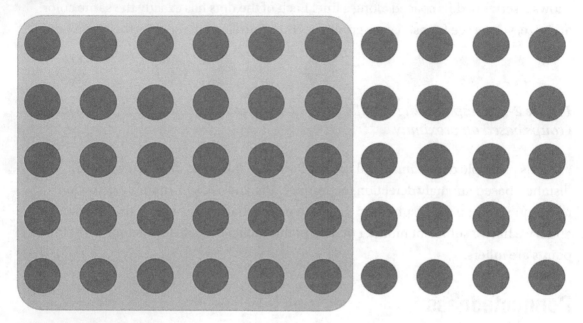

Figure 2-10. *Points within the gray region are part of one visual group, with points outside the space inhabiting an implicit second visual group*

Symmetry

We tend to perceive elements arranged symmetrically as part of the same group. This is the principle of *symmetry*. Although the principle is named symmetry, it applies as well to reflection, rotation, and translation. As an example of this, if I see a person standing in front of a mirror, my brain does not perceive the doppelgänger as a different person. As a more abstract example, I can rotate a triangle in any fashion and still recognize that it is a triangle. If I see two triangles, one that is rotated differently from the other, I still will group them together despite the rotational difference. Moving back to practical examples, a set of arrows can point the way through corridors, and despite subsequent arrows pointing in different directions—sometimes pointing to the left, sometimes pointing to the right, sometimes pointing straight ahead—I can still group them together as the set of directions I must follow in order to reach my destination.

One way to increase the likelihood of a person making a match based on symmetry is to use the properties of similarity. Going back to the prior example of arrows leading the way, it helps if all of the arrows have the same color and are approximately the same size. Doing so reduces the likelihood that a person believes arrows of one color belong to a set of directions and arrows of a different color belong to a separate set of directions. This strategy of arrows leading the way is common in hospitals, which use arrows and colored lines on walls to indicate the directions of different wards.

Common Fate

The principle of *common fate* informs that we tend to group together elements moving in the same direction. For example, if we look at an image of a roadway filled with cars headed in one direction, we think of this as a grouped noun: traffic. The individual vehicles become part of a single mental group, but that group does not include cars headed in the opposite direction in the same image.

We can tie this to the visualization world by looking at time series data. The closer two series appear to move in coordination, the more we think of them as a single group. Figure 2-11 shows an example of Consumer Price Index in the category of apparel for three regions of the United States between January of 1997 and July of 2021.

Consumer Price Index for Apparel by Region

Region ● Midwest ● South ● West

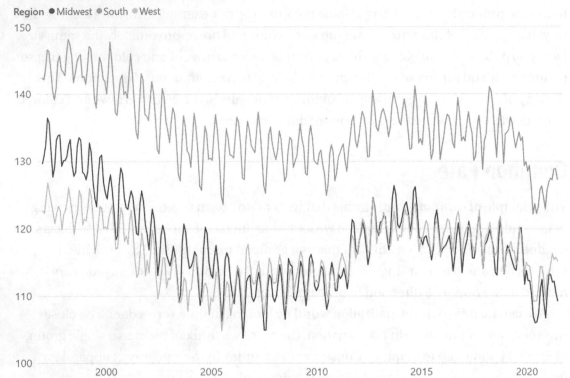

Figure 2-11. *Three regions follow a similar pattern*

Each series follows a common pattern, although the Midwest series has a steeper decline in the early years. Because the patterns are much the same, with similar ups and downs, we interpret the three lines as following one common pattern and can speak of "the" pattern with respect to the three separate lines.

Synchrony

The final principle we will cover is *synchrony*. In this case, we group together visual elements that represent a synchronous event. Another way of stating this is that multiple things happening at approximately the same time will seem to be related. Suppose I look out my window and see a car speeding by, followed closely on by a police cruiser with lights flashing and siren blaring. I will naturally assume that these two events are related to one another and may be able to construct in my mind a story that explains how the situation came about.

When thinking about synchrony, however, the notion that "correlation does not imply causation" is important to keep in mind. The fact that two things happen in temporal proximity does not mean that they are related. The two cars speeding along the road may actually be unrelated, as perhaps the police cruiser is speeding toward a major incident and does not even have time to report or care about the other speeding vehicle. In that case, the two events occurring in near order is just happenstance.

Synchrony and common fate both play a strong role in time series analysis, which will be the focus of Part IV.

Helping People Find Anomalies

We can use Gestalt principles to help our end users find anomalies when displaying data in a visual format. In the following section, we will summarize key techniques you can use to help report viewers and application users spot outliers and anomalies.

Use Color As a Signal

Our ability to view differences in color is one of the most important signal indicators we have as humans. Even for individuals with various forms of color vision deficiency (colloquially referred to as color blindness), spotting a difference in color is something our minds do without our needing to pay attention to the fact. Color is an emergent phenomenon, to use Gestalt terminology.

Given this, we should limit the use of color changes to cases in which it makes the most sense. If we create a visual that looks like a rainbow exploded on it, the sheer number of color changes will overwhelm our brain's limited ability effectively to process changes. As a result, we introduce a source of noise in our image. This noise may not be exactly the same as what we defined in Chapter 1 (in which noise is an outlier that is not interesting), but it is the same in that it serves as a distraction to the viewer.

Instead of splattering color across the visual, take care and use color changes to represent matters of significance. By limiting the number of colors you use on a visual, it will be possible for viewers to spot outliers, allowing a thoughtful viewer with sufficient domain knowledge the ability to investigate and determine the cause of outliers.

Limit Nonmeaningful Information

As an analog to our first point, we want to limit nonmeaningful information in visuals. The goal is to keep the important part of the story as the figure rather than the ground. That said, the larger the number of things competing for attention in a visual, the more likely some superfluous, non-informative element is to take away our focus. We've seen in these principles that humans are adept at picking out differences and spotting outliers, but we have similarly seen that the same principles that work in our favor can also act to trick us into missing something that is there or seeing something that isn't.

Enable "Connecting the Dots"

Finally, display things that allow viewers to "connect the dots" and apply their domain knowledge in the most effective way. The best way to enable connecting the dots is to take advantage of principles such as proximity, similarity, and common fate.

When it comes to proximity, try to lay out visuals in such a way that points that are near one another are close in relevant ways. This makes intuitive sense when plotting dots on a number line, as dots close to one another will be similar in number. In an ideal world, you would be able to set the scale such that dots that appear close to a viewer are part of the same relevant cluster and that separations between dots represent significant differences. You can't always control this perfectly but think about the size of the image, what your minimum and maximum axis values are, and how big the dots are.

To improve viewers' ability to use the principle of similarity, try to use common shapes, colors, and sizes to represent the same things. If you do have shapes of different sizes, ensure that the size represents a meaningful continuous variable, such as number of incidents. Also, try to ensure that the relative differences in scale make sense so that you don't have one dot that completely overwhelms all of the rest.

Finally, with respect to common fate, try, if you can, to emphasize or point out cases in which things move similarly, grouping them in a way separate from the nonconforming elements. For example, reviewing Figure 2-11, we can see that three regions moved very similarly with respect to Consumer Price Index in the apparel category. It turns out that there is a fourth region whose behavior was a bit different. Figure 2-12 shows an example of using the same color to represent regions with a common fate and highlighting the Northeast region with a different color.

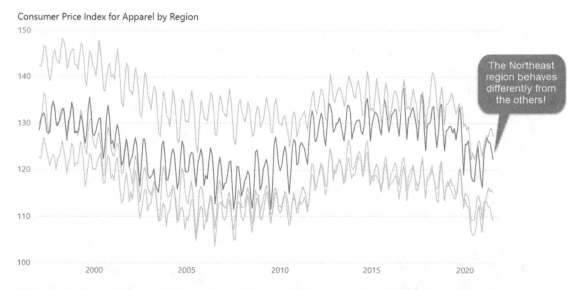

Figure 2-12. *Usage of the same color signifies regions with a common fate, allowing us to point out a region exhibiting somewhat different behavior*

Conclusion

More than a hundred years after Max Wertheimer's seminal paper, the Gestalt school remains relevant, especially in the field of sensory recognition. Having a solid understanding of the principles laid out in this chapter will greatly benefit report viewers and application users aiming to make sense of the information you provide. This is a laudable goal, but when it comes to anomaly detection, it is not sufficient to the task of creating a program that can allow a computer to find outliers in data. Even so, we will come back to these principles several times throughout the book, as ultimately, the problem of anomaly detection is a human endeavor: it is humans who provide the necessary criterion of meaningfulness to outliers in order to separate anomalies from mere noise.

## CHAPTER 3

Formalizing Anomaly Detection

In the prior chapter, we looked at the sensory tools humans have to perform outlier and anomaly detection, taking the time to understand how they work. Computers have none of these sensory tools, and so we need to use other processes to allow them to detect outliers. Specifically, we need to implement at least one of the three approaches to outlier detection in order to give a computer some process for outlier detection. This chapter will begin to lay out the first of the three approaches to outlier detection that we will cover throughout the book: the statistical approach.

The Importance of Formalization

Chapter 2 made it clear that humans are, on the whole, very good at pattern matching and identifying problems. Many of the skills humans have used to survive in unforgiving ecosystems can also be helpful in observing more prosaic things, such as corporate scatter plots of revenue by business unit over time. That said, humans are fallible, as are the tools they use. In this section, we will look at several reasons why an entirely human-driven outlier detection program is unlikely to succeed.

"I'll Know It When I See It" Isn't Enough

Making a set of raw data available to a person and expecting to get a complete and accurate set of outliers from it is typically a fool's errand. Our processing capabilities are paradoxical: in some circumstances, we can easily discern a pattern or find an oddity in a sea of information, but even small datasets can be overwhelming to our brains without formal tools in place. Furthermore, many of the Gestalt insights we covered in the prior chapter have a learned aspect to them. For example, emergence occurs when

© Kevin Feasel 2022
K. Feasel, *Finding Ghosts in Your Data*, https://doi.org/10.1007/978-1-4842-8870-2_3

the brain *recognizes* a pattern, meaningfulness applies to *personally relevant* scenes, and reification depends on our *recognizing* a thing. Two people who have different backgrounds and experiences may not find the same outliers in a given dataset for that reason. Because we cannot assume that all viewers will have the same backgrounds, experiences, and capabilities, handing data to viewers and expecting outputs is untenable.

Human Fallibility

This also leads into the next point: humans make mistakes. Our eyes are not perfect, and we do not always perfectly recall specific values, patterns, or concepts. Because we have a tendency to forget prior patterns, observe spurious or nonexistent patterns, and not observe patterns right in front of our faces, we aren't perfectly reliable instruments for outlier detection.

I don't want to oversell this point, however—humans *are* very good at the task and should be an important part of any anomaly detection system. At the very least, humans are necessary in order to convey significance to outliers and determine whether a given outlier is ultimately anomalous. Beyond that, humans can also provide some capability in finding outliers the machine or automated technique misses. Incidentally, this helps us get a feeling for how good an automated outlier detection process is—if humans are frequently finding anomalies that the process missed, that's not a good sign!

Marginal Outliers

One class of human fallibility is the case of the marginal outlier. When looking at a graph, certain types of outliers are easy to spot. Figure 3-1 is a reproduction of an image we saw in Chapter 1, and it is easy to spot the two outliers in the example.

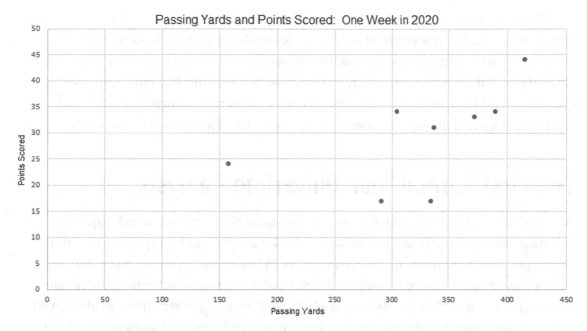

Figure 3-1. *Points scored vs. yards of passing for a single week in the NFL. We see two outliers on this chart, neither of which hit 20 points scored*

Those are easy to spot, but is the topmost point, in which a quarterback threw for approximately 415 yards and his team scored 44 points, an outlier? If not, what if his team scored 45 or 47 points? How many points would it take for that data point to become an outlier? Humans are great at spotting things that are quite different from the norm, but when it comes to noticing a very small difference, we tend not to be nearly as accurate. As a result, we can miss cases in which there was an outlier but it was just on the edge or add in cases that are actually inliers but are close to the edge.

The Limits of Visualization

Finally, humans do best when reviewing data in a visual form. Figures like 3-1 allow us to take advantage of our sensory perception capabilities much more efficiently than tables of numbers. Humans can do quite well managing a few variables, but the maximum number of variables the average person can deal with at a time is five: one each for the X, Y, and Z axes; one to represent the size of data points; and capturing a time measure by moving points across the chart in time. Even this may be pushing it in many circumstances, leaving us comfortable with no more than two or three variables.

What this tells us is that once things get complex and we need to look at relationships in higher and higher dimensional orders, Mark 1 eyeballs tend not to do so well. We can potentially use techniques such as dimensionality reduction to make the job easier for analysis, but even dimensionality reduction techniques like Principal Component Analysis (PCA) have their limits. In more complex scenarios, we may still not be able to get down to a small enough number of components to allow for easy visualization.

The First Formal Tool: Univariate Analysis

With the prior section in mind, we can introduce additional tools and techniques to enhance human capabilities and make it easier for subject matter experts to pick out anomalies from the crowd. In this chapter, we will begin with a discussion of statistical analysis techniques and specifically *univariate analysis*. The term "univariate" means "one variable" and refers to a dataset in which we have a single variable to analyze. And when we say "one variable," we mean it: it's simply a collection of values which we might implement using something like a one-dimensional array.

This approach may sound far too limiting for any realistic analysis, but in practice, it's actually quite effective. In many circumstances, we can narrow down our analysis to a single important input by assuming that any other variables are *independent*, meaning that no outside variable affects the value of a given data point. This is a naïve assumption that is almost never true, and yet it often works out well enough in practice that we can perform a reasonable statistical analysis regardless.

Distributions and Histograms

When looking at one-dimensional data, our focus is on the distribution of that data. A simple way to think of a *distribution* (or a *statistical distribution*) is the shape of the data. That is, how often do we see any particular value in the dataset? For one-dimensional data, we can create a plot called a *histogram*, which shows the frequency with which we see a value in our dataset.

Before we see a histogram in action, let's briefly cover one property of variables, which is whether the variable is *discrete* or *continuous*. A discrete variable has a finite and known number of possible values. If I ask you to give a movie a number of stars ranging from 1 to 5, with 5 being the highest and no partial stars allowed, I have created a discrete variable that contains a set of responses ranging from 1 to 5, inclusive.

By contrast, a continuous variable has an infinite number of possible values. In practice, though, "infinite" often means "a very large number" rather than truly infinite. If I asked a series of individuals how much money they made last year (translated to US dollars if necessary), I will get a wide variety of responses from people. Although there is a finite number of dollars and cents we could expect a single person to make in a single year—obviously, somewhere short of the combined GDP of all of the countries in the world last year—we consider this variable to be continuous in practice because we are likely to get a large number of unique responses.

Furthermore, when building a histogram from a continuous variable, we typically don't care as much about the exact value of the variable but can make do instead with an approximation, allowing us to aggregate the values within a certain range and display them as a single group, which we call a *bin*. Figure 3-2 shows an example of one such histogram, containing the Earned Run Average (ERA) of all pitchers in Major League Baseball who threw at least 150 innings in a given season, giving us 4897 such entries. I generated this image using the *ggplot2* library in R, which has a default of 30 bins.

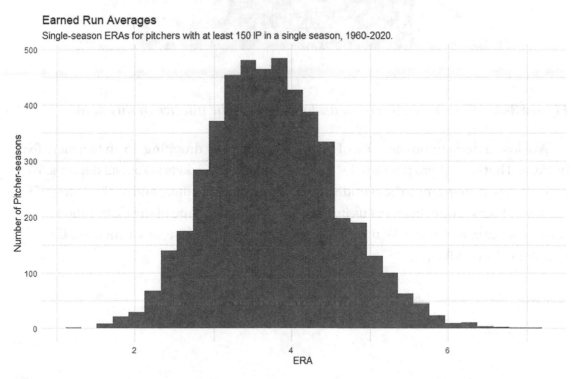

Figure 3-2. *A histogram of single-season Earned Run Averages using 30 bins*

Figure 3-3 shows the importance of bin selection, as we decrease the number of bins to 5.

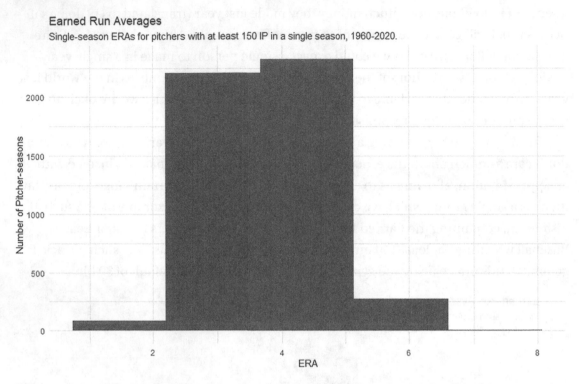

Figure 3-3. *When using five bins, we lose much of the nuance in this data*

We lose a tremendous amount of fidelity in the data by dropping down to a mere five buckets. That said, there is no single "correct" number of buckets across all datasets. We typically use histograms to help understand the shape of the data, and so the "correct" number of bins is one that gives us information about the shape of the distribution without an excess of noise. Figure 3-4 shows an example of noise, as we increase the number of bins to 500.

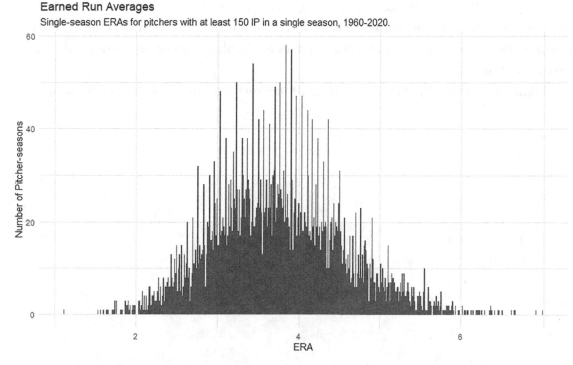

Figure 3-4. *Having too many bins can hinder our understanding of the data just as having too few bins does*

In this case, breaking the data into 500 bins hinders our understanding of its distribution; there are simply too many bins to make much sense of the resulting image. One rule of thumb when selecting bin counts is to decrease the number of bins until it stops looking spiky, as in Figure 3-4. With this example, that number is probably in the range of 30–50 bins, and the default was certainly good enough to give us an idea of the overall feel of our data.

Now that we have a basic idea of a distribution and know of one tool we can use to display the distribution of a univariate variable, let's briefly cover the most famous of all distributions, as well as some of its relevant properties for anomaly detection.

The Normal Distribution

The *normal* or *Gaussian* distribution is easily the distribution with the greatest amount of name recognition. This distribution is one of the first any student learns in a statistics course and for good reason: it is easy to understand and has several helpful properties, and we see it often enough in practice to be interesting. Although there are other distributions that occur more frequently in nature—such as the log-normal distribution—let's make like a statistics textbook and focus on the normal distribution first.

Figure 3-5 shows the *probability density function* (PDF) of the normal distribution.

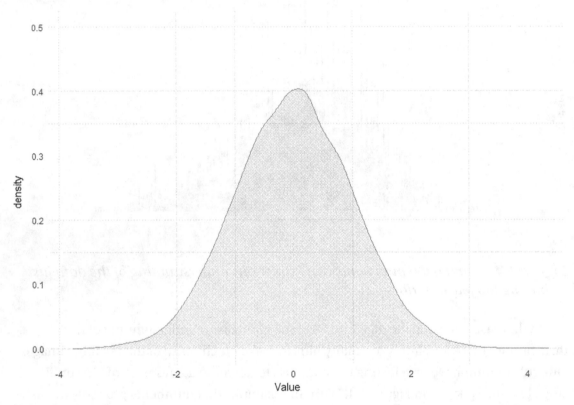

Figure 3-5. *The probability density function of a normal distribution. This PDF was generated by choosing 15,000 random values following a normal distribution and plotting the resulting density function*

This image shows us two things: first, what is the likelihood that if we were to pick a point at random from the normal distribution, we would end up with a specific value. The probability that we would select a specific value from an infinitesimal range along the X axis is its height on the Y axis. This tells us that we are more likely to end up with values close to 0 than values far away from 0. Second, we can add up the area under the curve and gain a picture of how likely it is we will select a value less than or equal to (or greater than or equal to) a specific value along the X axis. For example, if we were to sum up the area along the curve up to the value –2, this equates to approximately 2.5% of the total area of the curve. That tells us that if we randomly select some point from the normal distribution, we have approximately a 2.5% chance of selecting a point less than or equal to –2.

These two properties apply to almost all of the distributions we will look at in Chapter 3. Now that we have a basic idea of distributions, let's look at three measures that help us make sense of most distributions.

Mean, Variance, and Standard Deviation

The first measure of importance to us is the *mean*, one of the two terms (along with *median*) that we colloquially term "average." We can calculate the mean of a collection of points as $\mu = \dfrac{\sum values}{N}$. In other words, if we take the sum of values and divide by the count of values, we have the mean.

In the example of the normal distribution shown earlier, the mean is 0. Because the curve on a normal distribution is symmetrical and single peaked, the mean is easy to figure out from viewing the probability distribution function: it's the peak. With other distributions, the mean is not necessarily quite as easy to find, as we'll see in the following text.

Now that we have the mean figured out, the other key way of describing the normal distribution is the *variance*, or how spread out the curve is. The calculation of variance of a collection of points is a bit more complex than that of mean: $\sigma^2 = \dfrac{1}{N}\sum_{i=1}^{N}(x_i - \mu)^2$. In other words, we sum up the square of the differences from the mean. Then, we divide that by the number of data points, and this gives us our calculation of variance.

Once you understand variance, *standard deviation* is easy: standard deviation is the square root of the variance. Standard deviation is a rather important measure for a normal distribution because there are certain rules of thumb around how likely we are to randomly draw a point some distance from the mean. Figure 3-6 shows this 68 95 99.7 rule.

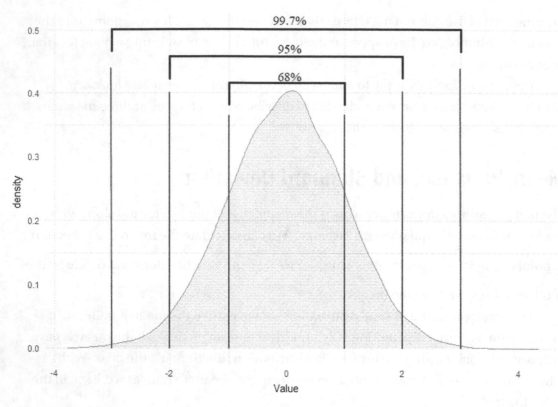

Figure 3-6. *The normal distribution follows the 68-95-99.7 rule*

As we can see from Figure 3-6, the likelihood of drawing a value between –1 and 1 is approximately 68%. There's a 95% chance we will draw a value between –2 and 2. Finally, there is approximately a 99.7% chance we will draw a value between –3 and 3. This leads us to our first technical definition of an outlier: a given point is an outlier if it is more than three standard deviations from the mean and we assume that the underlying distribution is a normal distribution. If the distribution is, in fact, normal, this means approximately 3 in 1000 points will fall outside the norm. The further from the mean a point is, the more likely it is to be anomalous. Suppose that we find a point six standard deviations from the mean. There is approximately a 1 in 500 million chance of finding a data point six standard deviations from the mean, so unless you process billions of rows of data, there's a pretty good chance this point six standard deviations from the mean didn't really come from our normal distribution and likely came from some other distribution instead. This is akin to saying, as we did in Chapter 1, that some other process likely generated that data point.

Before we close this section, let me differentiate *the* normal distribution from *a* normal distribution. The normal distribution is one with a mean of 0 and a variance of 1. Any other distribution that otherwise follows the properties of the normal distribution but has a mean other than 0 or variance other than 1 still operates like the normal distribution. The reason for that is we can perform equivalent operations on each point in our collection to get us to the normal distribution. What this means in practice is that we can apply the same set of tools to all of the distributions in Figure 3-7.

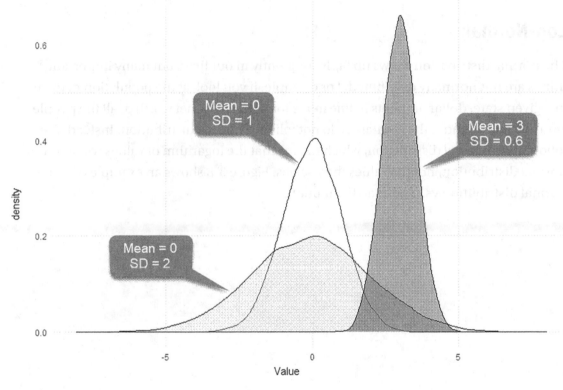

Figure 3-7. *Three normal distributions appear on this chart. The first normal distribution is the normal distribution, with a mean of 0 and standard deviation of 1. The second, squatter curve represents a normal distribution with a mean of 0 and standard deviation of 2. The third has a mean of 3 and standard deviation of 0.6*

When the variance is not 1, we can multiply each point by some scaling factor to make the variance 1. When the mean is not 0, we can add some factor (positive or negative) to make the mean 0. Importantly, these operations do not affect the relationships between elements in the data, which allows us in practice to treat a distribution as normal even if its mean or variance differs from that of the normal distribution.

Additional Distributions

The normal distribution is the first distribution we reviewed because it has several nice properties and shows up often enough in practice. It is, however, certainly not the only relevant distribution for us. Let's now look at three additional distributions, each of which adds its own nuances. Note that there are plenty of distributions we will not cover in this chapter due to time and scope. These distributions can still be important, so please do not consider this set comprehensive or covering all practical examples.

Log-Normal

The normal distribution shows up fairly frequently in our lives, but many important things are not normally distributed. For example, if you look at the population of towns in a given state, dollar amounts in line items for budgets, or net worth of all the people who live in a country, these datasets do not follow a normal distribution. Instead, they follow a *log-normal* distribution, which means that the logarithm of values will follow a normal distribution, not the values themselves. Figure 3-8 shows an example of a log-normal distribution vs. a normal distribution.

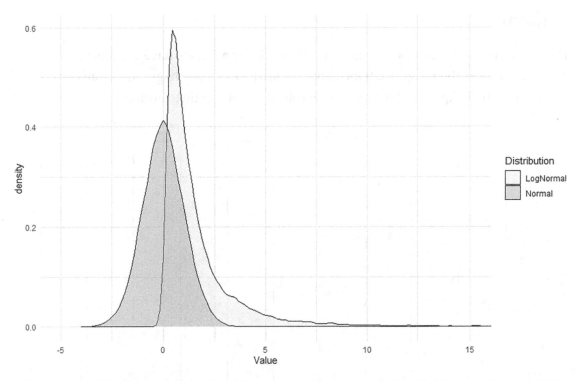

Figure 3-8. *The log-normal distribution vs. the normal distribution, using 15,000 sampled points from each distribution. The log-normal distribution exhibits a "long tail" effect. Note that the graphic is slightly inaccurate: the lowest log-normal value in this dataset is 0.02, so all values are above 0*

The normal distribution follows a symmetric "bell curve" shape, but the log-normal distribution has a "long tail," meaning that you are more likely to see a value on the left-hand side of the peak than on the right-hand side. Another way to describe this is having *skew*, where one side of the curve stretches out significantly further than the other.

One other point to note is that because this is a logarithmic calculation, all values in your collection must be greater than zero as you cannot take the logarithm of a negative number or zero. If you have a phenomenon that cannot be negative (such as income earned over the past year), this is a tell that the distribution is more likely to be log-normal than normal.

If you assume a collection follows a normal distribution but it actually follows a log-normal distribution, you are likely to end up with a larger number of high-value outliers that turn out to be noise: that is, some people really are that wealthy or some cities really are that large.

Uniform

The uniform distribution is the easiest one to understand. If we have a uniform distribution over the range 0 to 9 inclusive, we are just as likely to get the value 0 as we are 9 as we are 5. Figure 3-9 shows an example of a uniform distribution.

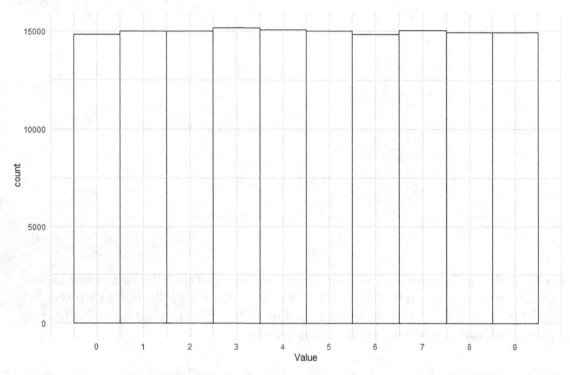

Figure 3-9. *150,000 points drawn from a uniform distribution ranging between 0 and 9. Because this is a sample of points, there are minor differences in counts that we would not expect to see in the ideal form of a uniform distribution*

With a uniform distribution, we cannot call any single point an outlier, as every value is equally likely to happen. Instead, if we have reason to believe that the distribution should be uniform—such as when using the RAND() function in a database like SQL Server or the random() function in Python—we are going to need to observe a sequence of events and calculate the likelihood of this sequence occurring given a uniform distribution. To put this in concrete terms, rolling a 6 on a six-sided die is pretty common—in fact, it should happen 1/6 of the time. But rolling 20 consecutive 6s should bring into question whether that die is rigged.

Cauchy

The final distribution is a glorious and thankfully uncommon one. The Cauchy distribution looks a lot like a normal distribution, as we can see in Figure 3-10.

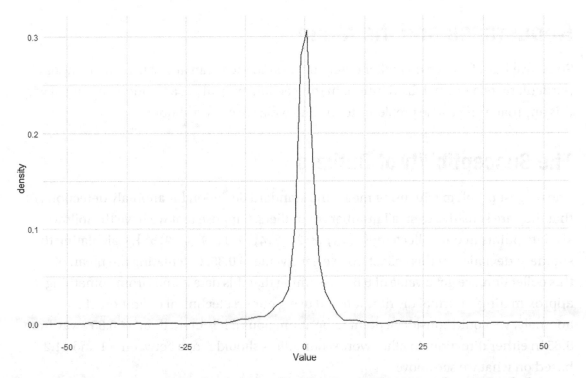

Figure 3-10. *An example of a Cauchy distribution with 1000 values sampled. The "fat tails" are evident in this image. In the sample itself, the minimum value was -425, and the maximum value was 222. By contrast, a normal distribution would fit between -4 and 4*

The key difference is that the Cauchy distribution has "fat tails" whereas the normal distribution does not. In other words, the normal distribution asymptotically approaches zero, meaning that you can realistically discount the likelihood of finding a point six or seven standard deviations from the mean and the further out you go, the more you can discount the likelihood. For a Cauchy distribution, however, there is always a realistic possibility of getting a point that looks 10 or 20 standard deviations from what you think is the mean. I needed to phrase the last sentence the way I did because a Cauchy distribution does not actually have a mean or variance. There is a point of highest probability that might look like a mean when looking at the probability distribution function, but because of the fat tails, we will never converge on a stable mean.

There are a few cases in which we see Cauchy distributions in nature, primarily in the field of spectroscopy. Fortunately, these occurrences are rare, as they make our job of figuring out whether something is an anomaly much harder!

Robustness and the Mean

So far, we have focused on outlier detection around the mean and standard deviation, particularly of a normal distribution. In this section, we will get an understanding of why this approach can cause problems for us and what we can do about it.

The Susceptibility of Outliers

The biggest problem with using mean and standard deviation for anomaly detection is that they are sensitive to small numbers of outliers. Suppose that we have the following six data points in our collection: { 7.3, 8.2, 8.4, 9.1, 9.3, 9.6 }. Calculating the standard deviation of this collection, we get a value of 0.85. Calculating the mean of this collection, we get a value of 8.65. Assuming that this data comes from something approximating a normal distribution and using our first technical definition of an anomaly, we could consider a point to be anomalous if it is more than 3 * 0.85 away from 8.65 in either direction. In other words, our inliers should range between 6.1 and 11.2, based on what we see above.

Let's now add one outlier value: 1.9. We can tell it is an outlier because it is well outside our expected range; thus, the system works. But now look what it does to our system: the standard deviation when you add in the value 1.9 becomes 2.67, and the mean drops to 7.69. Our range of acceptable values now becomes –0.32 up to 15.7, meaning that the item we just called an outlier would now become an inlier if it were to appear a second time!

The Median and "Robust" Statistics

This problem with mean and standard deviation led to the development of "robust" statistics, where robustness is defined as protecting against a certain number of outliers before the solution breaks down like the example shown previously. When calculating averages, we can switch from the nonrobust mean to the robust *median*. The median is the 50th percentile of a dataset, otherwise known as the midpoint of all values.

In our previous example, the median of the set { 7.3, 8.2, 8.4, 9.1, 9.3, 9.6 } is 8.75. To determine this, we remove the smallest and largest values as pairs until we are left with either one or two remaining values. In this case, we remove 7.3 and 9.6, then 8.2 and 9.3. That leaves us with 8.4 and 9.1. Because we have two values remaining, we take the midpoint of those two values, which is 8.75.

From there, we can calculate *Median Absolute Deviation*, or MAD. MAD is another robust statistic, one that acts as a replacement for standard deviation. The formula for this is $MAD = median\left(\left|X_i - \tilde{X}\right|\right)$. That is, we take the median of the absolute value of differences between individual values and the median of our set. Because our median is 8.75, we subtract 8.75 from each of the collection items and take its absolute value, leaving us with { 1.45, 0.55, 0.35, 0.35, 0.55, 0.85 }. The median of this collection is 0.55.

Applying a similar formula to before, we can call an outlier something that is more than three MAD away from the median. That is, if a value is more than 3 * 0.55 units distant from 8.75 (or outside the range of 7.1 to 10.4), we label the value an outlier. The value of 1.9 would certainly be an outlier here, but let's see what it does to our next calculation.

The median of the set { 1.9, 7.3, 8.2, 8.4, 9.1, 9.3, 9.6 } is 8.4. This already lets us see the concept of robustness in action: we have an outlier and drop it immediately. It does push the median down but only to the next inlier point. Then, to calculate MAD, we need the median of the absolute value of differences from 8.4, which is { 6.5, 1.1, 0.2, 0, 0.7, 0.9, 1.2 }. Determining the median of this collection gives us a value of 0.9 for the MAD. This means that our range of acceptable values has expanded to 3 * 0.9 from 8.4, or 5.7 to 11.1. There is some expansion but not nearly as much as before—we would still catch 1.9 as an outlier even with this expanded set. Furthermore, MAD "recovers" from an outlier much faster than standard deviation does, meaning that our window of acceptable values opens more slowly and closes more quickly with a robust statistic than with a nonrobust statistic, making robust statistics much less likely to "hide" outliers if we happen to have several in our dataset.

Beyond the Median: Calculating Percentiles

We have already looked at one calculation of percentile: the median, or 50th percentile. There are other percentiles that we find useful in outlier detection, and one of the most useful visuals contains several of them: the box plot. Figure 3-11 shows an example of a box plot.

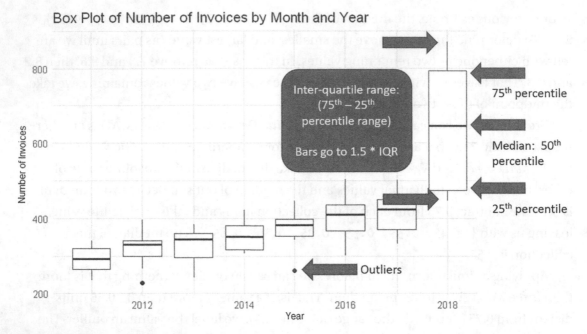

Figure 3-11. *An example of a box plot*

A box plot contains, at a minimum, five useful percentiles: the minimum value (0th percentile), maximum value (100th), median (50th percentile), and the 25th and 75th percentiles. The box itself is made up of the range between the 25th and 75th percentiles, which we call the *interquartile range* (IQR). We can see a line where the median is, giving us an idea of how skewed the data is (because data skewed in one direction will show a difference in size between the median and the outer percentile compared to its counterpart in the opposite direction). From there, we see "whiskers" stretching out from the box. The whiskers represent all data points within 1.5 * IQR units of the top or bottom of the box. Any points beyond this are outliers, giving us our third technical definition of an outlier: an outlier is any point more than 1.5 times the interquartile range below the 25th percentile or above the 75th percentile.

As we begin to collect some of these technical definitions of outliers, it is important to note that these definitions are not guaranteed to overlap, meaning that one process may call something an outlier but a second process may call it an inlier. We will deal with this complexity starting in Chapter 6.

Control Charts

Before we close this chapter, let's take a look at one more variant on visualizing anomalies: the *statistical process control chart*, which I'll refer to as a control chart. The main purpose of a control chart is to determine whether a given process is working as expected. One popular use case for control charts is in the manufacturing world, in which operations have certain allowable tolerances. For example, suppose we wish to monitor a machine that rivets together two pieces of sheet metal. We know the expected positions of the rivets and also know the maximum variance from that expected position we can allow before we need to fix or scrap the sheet metal. With this information, we can create a control chart like the one in Figure 3-12. This particular kind of chart is called an I chart and measures individual values at regular intervals, such as the distance of a rivet from the "perfect" position. We track these values over time as the machine processes more data and compare each value vs. the mean.

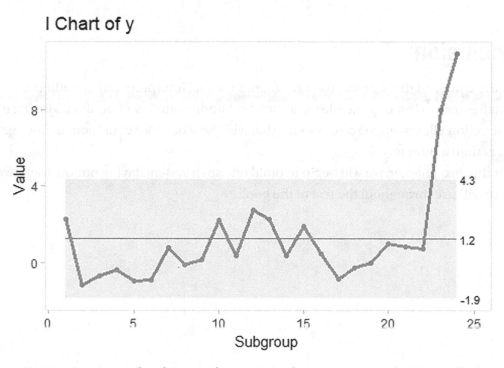

Figure 3-12. *An example of a specific statistical process control chart called an I chart. The shaded box represents three standard deviations from the line marking the mean. The last two data points are outside the box, indicating that this process has just gone "out of control"*

In this chart, we see the expected value as a line and the tolerable range as a shaded box. Within the United States, regulated industries that require control charts typically set the limits to three standard deviations on each side of the mean, and this makes up the shaded area in the figure. This might give you pause when thinking about the problem of standard deviation being a nonrobust statistic, but there is one key difference here: with a control chart, we have pre-calculated the mean and standard deviation and do not update it based on our new values. Therefore, if new information comes in indicating outliers, it does not affect the position or size of the shaded area.

We can see from this example that the process has started out "in control," meaning that it is within the shaded area. After some number of steps, it has gone "out of control" and indicates that there is a problem. Using our prior example, we might see that the rivets are too close together or that they are too far to the left or right of where they should be. Whatever the cause may be, the control chart tells us that we need to take action and resolve this problem.

Conclusion

Over the course of this chapter, we moved from a visual interpretation of outliers and anomalies using Gestalt principles to a statistical understanding of outliers. Moving in this direction allows us to create systems that will likely be more consistent at finding outliers than a human.

In the next chapter, we will begin to build one such system, laying out the framework that we will use throughout the rest of the book.

PART II

Building an Anomaly Detector

In this second part of the book, we will build the framework for an anomaly detection engine and flesh out one of the four use cases: univariate anomaly detection. We begin in Chapter 4 with an overview of tools and tech stack choices, comparing alternatives and making implementation decisions. Then, we will lay out the basic framework we will use for the rest of the book. Chapter 5 picks up where Chapter 4 left off, enabling us to create unit tests along the way. With our framework in place, we implement our first detection techniques (and our first ensemble) in Chapter 6. We extend this out in Chapter 7, adding a variety of tests for normally distributed data. Finally, in Chapter 8, we build a simple web application to visualize our results.

Laying Out the Framework

To this point, we have focused entirely on the theoretical aspects of outlier and anomaly detection. We will still need to delve into theory on several other occasions in later chapters, but we have enough to get started on developing a proper anomaly detection service.

In this chapter, we will build the scaffolding for a real-time outlier detection service. Then, in the next chapter, we will integrate a testing library to reduce the risk of breaking our code as we iterate on techniques.

Tools of the Trade

Before we write a line of code, we will need to make several important decisions around programming languages, which packages we want to use to make the process easier, and even how we want to interact with end users and other programs. This last choice includes decisions around protocols and how we wish to *serialize* and *deserialize* our data to ensure that calling our service is as easy as possible while still allowing us to deliver on our promises.

Choosing a Programming Language

The first major choice involves picking a programming language. For this book, we will use the Python programming language to work with our service. Python is an extremely popular programming language, both as a general-purpose language and especially in the fields of data science and machine learning. In particular, we will use the Anaconda distribution of Python, as it comes with a variety of useful libraries for our anomaly detection project. This includes pandas, which offers table-like data frames; numpy, a library that provides a variety of mathematical and statistical functions; and scikit-learn, a machine learning library for Python. In addition to these libraries, we will introduce and use several more throughout the course of this book.

© Kevin Feasel 2022
K. Feasel, *Finding Ghosts in Your Data*, https://doi.org/10.1007/978-1-4842-8870-2_4

Python is, of course, not the only programming language that you can use to build an anomaly detection service. Several other languages have built-in functionality or easy-to-install libraries to make anomaly detection a fairly straightforward process. For example, the R programming language includes a variety of anomaly detection packages, including the excellent `anomalydetector` library. R is an outstanding domain-specific language for data science, data analysis, and statistics. Although it is not necessarily a great general-purpose programming language, what we want is well within its capabilities, and the author and technical editor have previously combined to build a fully featured anomaly detection service in R as well.

If you wish to stick to general-purpose programming languages, the .NET Framework has two such languages that would also be quite suitable for an anomaly detection engine: the object-oriented C# and functional F# languages both can work with a variety of packages to make statistical analysis easier. The `Math.NET` project, for example, includes several packages intended for solving problems around combinatorics, numerical analysis, and statistical analysis.

In short, we will use Python because it is both a good choice and a popular choice for this sort of development, but do not feel obligated to use only Python for this task.

Making Plumbing Choices

Now that we have our choice of language covered, we should make a few decisions about how our end users will interact with the service. The first question we should ask is, why do our customers need this service? The more we understand what kinds of problems they want to solve, the better we can tailor our service to fit their needs. For our scenario, let's suppose that our end users wish to call our service and get a response back in near real time, rather than sending in large datasets and requesting batch operations over the entire dataset. The techniques we will use throughout the book will work just as well for both situations, but how we expose our service to the outside world will differ dramatically as a result of this decision.

Given that we wish to allow end users to interact in a real-time scenario, we will likely wish to use *HyperText Transfer Protocol* (HTTP) as our transfer mechanism of choice. We could alternatively design our own solution using *Transmission Control Protocol* (TCP) sockets, but this would add a significant amount of development overhead to creating a solution and would require that our users develop custom clients to interact with our service. Developing your own custom TCP solution could result in better performance but is well outside the scope of this book.

Sticking with HTTP as our protocol of choice, we now have two primary options for how we create services, both of which have a significant amount of support in Python: the Remote Procedure Call framework gRPC or building a *Representational State Transfer* (REST) API using *JavaScript Object Notation* (JSON) to pass data back and forth between our service and the client. The gRPC-based solution has some significant advantages, starting with payload size. Payloads in gRPC are in the *Protocol Buffer* (*Protobuf*) format, a binary format that compacts down request sizes pretty well. By contrast, JSON is a relatively large, uncompressed text format. JSON is more compact than other formats like *Extensible Markup Language* (XML), but there can be a significant difference in payload size between Protobuf and JSON, especially in batch processing scenarios. We also have strict contracts when working with gRPC, meaning that clients know exactly what the server will provide, what parameters and flags exist, and how to form requests. This can be a significant advantage over earlier implementations of REST APIs, though more recent REST APIs implement the OpenAPI specification (formerly known as *Swagger*), which describes REST APIs. The OpenAPI specification provides similar information to what gRPC describes in its contracts, making cross-service development significantly easier.

For the purposes of this book, we will choose to work with a REST API using JSON. There are three reasons for this choice. First, working with REST allows us to use freely available tools to interact with our service, including web browsers. The gRPC framework has very limited support for browser-based interaction, whereas REST works natively with browsers. The second reason for choosing REST over gRPC is that a REST API with JSON support provides us human-readable requests, making it easier for humans to parse and interpret requests and correct potential bugs in the client or the service. There are techniques to translate Protobuf requests to JSON, but for educational purposes, we will stick with the slower and easier-to-understand method. The final reason we will use REST APIs over gRPC is that more developers are familiar with the former than the latter. The move toward microservice-based development has made gRPC a popular framework, but we in the industry have several decades worth of experience with REST and have sorted out most of its problems. If you are developing an anomaly detection suite for production release, however, gRPC would be a great choice, especially if you develop the client or callers are familiar with the framework.

Reducing Architectural Variables

Now that we have landed on some of the "plumbing" decisions, we can look for other ways to simplify the development process. For example, considering that we intend to implement a REST API for callers, there are several Python-based frameworks to make API development easy, including Flask and FastAPI. Flask is a venerable package for core API development and integrates well with proxy solutions like Gunicorn. FastAPI, meanwhile, is a newer solution that has a lot going for it, including great performance and automatic implementation of the OpenAPI specification using a tool called ReDoc. Figure 4-1 shows an example of documentation built from the API we will develop in this chapter.

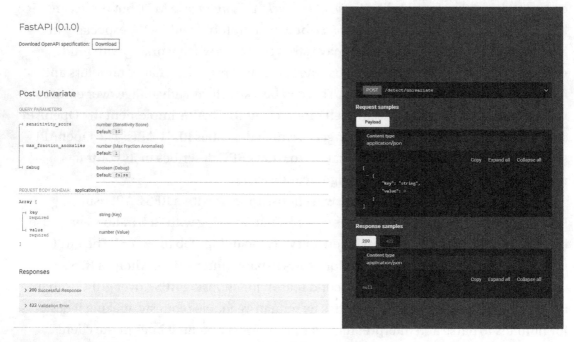

Figure 4-1. *Automatic API documentation for our API*

Beyond this, we can also use Docker containers to deploy the application. That way, no matter which operating system you are running or what libraries you have installed, you can easily deploy the end solution. Using the container version of the solution we build is optional, but if you do have Docker installed and want to simplify the dependency management process, there are notes for Docker-based deployment at the end of the chapter.

Developing an Initial Framework

After deciding on languages and technologies, it's time to begin development. In this section, we will install prerequisites for hosting the API. Then, we will create a stub API. In the next section, we will begin filling in the details.

Battlespace Preparation

You will need to have Python 3.5 or later installed on your computer if you wish to follow along and create your own outlier detection API service. The easiest way to get started is to install the Anaconda distribution of Python at `https://anaconda.com`. From there, install the Anaconda Individual Edition, which is available for Windows, macOS, and Linux. Anaconda also comes with a variety of Python libraries preinstalled, making it a great option for data science and machine learning operations.

Next, if you have not already, grab the code repository for this book at `https://github.com/Apress/finding-ghosts-in-your-data` and follow the instructions in `\src\README.md`, depending on whether you wish to follow along with the code in this book and create your own outlier detector or if you wish to use the completed version of the code base. Note that if you wish to follow along with the code in this book, you should still reference the completed version of the code, as certain segments of code will be elided for the purpose of saving space.

The repository includes several folders. The `\doc\` folder includes basic documentation about the project, including how to run different parts of the project. The `\src\app\` folder includes the completed version of the code base we will work throughout the book to create. The `\src\comp\` folder includes comparison work we will cover in the final chapter of this book. The `\src\web\` folder includes a companion website we will create and update throughout the book. Leaving the `\src\` directory altogether, the `\test\` folder features two separate sets of tests: one set of unit tests and one set of integration tests. Finally, we have a `Dockerfile` for people who wish to run the solution as a Docker container, as well as a `requirements.txt` file.

After installing Anaconda (or some similar variant of Python), open the `requirements.txt` file in the code repository. This file contains a set of packages, one per line, which we will use throughout the course of this book. You can install these packages manually using the `pip` packaging system, or you can open a shell in the base of the repository and run `pip install -r requirements.txt`.

Tip   If you receive an error message indicating that pip cannot be found, make sure that your Path environment variable includes Anaconda and its directories. For example, if you installed Anaconda at `E:\Anaconda3`, include that directory as well as `E:\Anaconda3\Library\bin`, `E:\Anaconda3\Library\usr\bin\`, and `E:\Anaconda\Scripts` to your path. Then, restart your shell and try again.

If you are using Anaconda, you will want to ensure that you are working under a new conda environment. You can create an environment in conda with the command `conda create --name finding_ghosts` inside the main directory of the code repository. Then, activate the environment with `conda activate finding_ghosts`. From there, you can run pip commands without affecting other installations of packages.

Framing the API

Now that we have everything installed, let's build an API. If you have retrieved the source code from the accompanying repository, be sure to rename the existing app folder to something like app_complete. Then, create a new app folder. Inside this folder, create an empty Python file named __init__.py, ensuring that you have two underscores before and after the word "init." This file helps by letting the Python interpreter know that our app folder contains code for a Python module. After creating this file, create another file called main.py. This file will be the entry point for our API. Open this new file with the editor of your choice. If you do not have an editor of choice, three good options are Visual Studio Code, Wing IDE, and PyCharm. Anaconda also comes with Spyder, another integrated development environment (IDE) for Python.

At the top of main.py, enter the code from Listing 4-1. This will ensure that we can reference the projects and libraries we will need for our API.

Listing 4-1. Import statements in the main.py file

```
from fastapi import FastAPI
from pydantic import BaseModel
from typing import Optional, List
import pandas as pd
import json
import datetime
```

The first three entries relate to the FastAPI library, allowing us to set up an API server, create models to shape our input and output signatures, and allow us to tag specific model elements as optional or as lists of items coming in. The next three entries relate to additional Python libraries that will be helpful. Pandas is a library for data analysis, and one of its most useful features is the ability to create Pandas DataFrames. DataFrames behave similarly to their counterparts in other languages and to tables in SQL, in that they are two-dimensional data structures with named entities, where each entity may have a different type. In addition to Pandas, we import the JSON and datetime libraries to make available their respective functionality.

Now that we have the import statements in place, the next step is to create a new FastAPI application and try it out. Listing 4-2 provides the minimal amount of code needed to build a FastAPI service.

Listing 4-2. Create a new API with one endpoint

```
app = FastAPI()

@app.get("/")
def doc():
  return {
    "message": "Welcome to the anomaly detector service, based on the book
    Finding Ghosts in Your Data!",
    "documentation": "If you want to see the OpenAPI specification,
    navigate to the /redoc/ path on this server."
  }
```

In this block of code, we first create a new API service called app. We then create a new endpoint for our application at the root directory. This endpoint accepts the GET method in REST. The GET method does not take a request body, and it is expected to return something. When navigating to web pages or clicking links using a browser, the browser translates these statements to GET operations.

After defining the endpoint, we now need to write out what our code should do if someone were to call that endpoint. In this case, we return a message. The message must be valid JSON. In this case, we create a JSON object with two attributes: message and documentation. The documentation attribute points us toward a path /redoc/, meaning that if you are running this on your local machine, you would navigate to http://127.0.0.1/redoc to see an OpenAPI specification similar to that in Figure 4-1.

In order to see this listing, navigate to the /src/ folder in a shell and enter the following command:

```
uvicorn app.main:app --host 0.0.0.0 --port 80
```

This command executes the uvicorn web server. It will then look for a folder called app and a file called main.py. Inside that file, uvicorn will find our reference to app and make the API available. We will run this on the local host on port 80, although you can change the host and port as necessary. Navigating to http://localhost will return a JSON snippet similar to that in Figure 4-2.

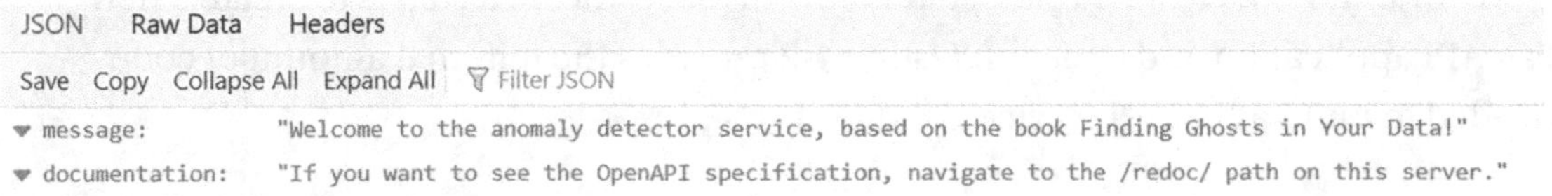

Figure 4-2. *The JSON message we get upon performing a GET operation on the /endpoint of the anomaly detector service. Some browsers, like Firefox, will generate an aesthetically pleasing JSON result like you see here; others will simply lay out the JSON as text on the screen*

Now that we have a functional API service, the next step is to lay out the input and output signatures we will need throughout the book.

Input and Output Signatures

Each API endpoint will need its own specific input and output structure. In this section, we will lay out the primary endpoints, define their signatures, and create *stub methods* for each. Stub methods allow us to define the structure of the API without committing to all of the implementation details. They will also allow us to tackle one problem at a time during development while still making clear progress.

In all, we will have four methods, each of which will correspond to several chapters in the book. The first method will detect univariate data using statistical techniques, the project of Part II. The second method will cover multivariate anomaly detection using both clustering and nonclustering techniques, which is Part III. The third method will allow us to perform time series analysis on a single stream of data, the first goal of Part IV. The final method will let us detect anomalies in multiple time series datasets, which we will cover in the second half of Part IV.

Defining a Common Signature

At this point, it makes sense to frame out one of our stub methods and see if we can design something that works reasonably well across all of our inputs. We will start with the simplest case: univariate, unordered, numeric, non-time series data. Each change from there will introduce its own complexities, but there should be a fairly limited modification at the API level, leaving most of that complexity to deeper parts of the solution.

All of the API calls we will need to make will follow a common signature, as defined in Listing 4-3.

Listing 4.3. The shell of a common solution

```
@app.post("/detect/univariate")
def post_univariate(
  input_data: List[<input class>],
  debug: bool = False,
  <other inputs>
):
  df = pd.DataFrame(i.__dict__ for i in input_data)

  (df, ...) = univariate.detect_univariate_statistical(df, ...)

  results = { "anomalies": json.loads(df.to_json(orient='records')) }

  if (debug):
    # TODO: add debug data
    results.update({ "debug_msg": "This is a logging message." })
  return results
```

We first need to define that we will make a POST call to the service. This method, unlike GET, does accept a request body and will allow us to pass in a dataset in JSON format. The post call will also define the endpoint we need to call. In this case, we will call the /detect/univariate endpoint, which will trigger a call to the post_univariate() method. This method will look very similar for each of the four cases.

The first thing we will do is create a Pandas DataFrame from our JSON input. Putting the data into a DataFrame will make it much easier for us to operate on our input data, and many of the analysis techniques we will use throughout the book rely on data being in DataFrames.

After creating a DataFrame, we will call a detection method. In this case, the function is called `detect_univariate_statistical()` and will return a new DataFrame that includes the original data, along with an additional attribute describing whether that result is anomalous. We can then reshape this data as JSON, which is important because the outside world runs on JSON, not Pandas DataFrames.

We'll also include a debug section, which will help during development and include troubleshooting information. For now, we will leave this as a stub with some basic details, but as we flesh out exactly how the processes work, we will have a better understanding of what we can and should include here. These debug results will be in the same output JSON as our set of anomalies, meaning that our caller will not need any additional software or capabilities to read debug information.

Finally, we return the results JSON object. This same basic layout will serve us well for the other three methods.

Defining an Outlier

We now have defined most of the base methods for detecting outliers, but there are a couple of blocks we still need to fill out, particularly around inputs. As we develop methods, we will want to offer users some level of control concerning what constitutes an outlier. It is important to note, however, that we do not expect users to know the exact criteria that would define an outlier—if they did, they wouldn't need an outlier detector in the first place! There are, however, a couple of measures we can expect a user to control.

Sensitivity and Fraction of Anomalies

The first key measure is sensitivity. In this case, we do not mean sensitivity as a technical definition but rather a sliding scale. Ideally, callers will be able to control—without knowing exact details about thresholds—the likelihood that our outlier detector will flag something as an outlier. We will have the sensitivity score range from 1 to 100 inclusive, where 1 is least sensitive and 100 is most sensitive. How, exactly, we implement this will depend on the technique, but this gives callers a simple mechanism for the purpose.

Additionally, we will need to control the maximum fraction of anomalies we would expect to see in the dataset. Some outlier detection techniques expect an input that includes the maximum number of items that could be anomalous. Rather than trying to estimate this for our users, we can ask directly for this answer. This will range from 0.0 to 1.0 inclusive, where 1.0 means that every data point could potentially be an outlier. In practice, we would expect values ranging between 0.05 and 0.10, meaning no more than 5–10% of records are outliers. This fraction can also give us a hard limit on how many items we mark as outliers, meaning that if a caller sends in 100 data points and a max fraction of anomalies of 0.1, we guarantee we will return no more than ten outliers. This will work in conjunction with the sensitivity score: think of sensitivity score as a sliding scale and max fraction of anomalies as a hard cap.

Single Solution

If, for a given class of problem, there is a single best technique to solve the problem, we can use this technique to the exclusion of any other possible techniques. In that case, calculations are fairly straightforward: we apply the input data to the given algorithm, sending in the sensitivity score and max fraction of anomalies inputs if the technique calls for either (or both). We get back the number of outlier items and then need to determine whether we can simply send back this result set or if we need to perform additional work. For cases in which the input technique accepts sensitivity score or max fraction of anomalies, our work is probably complete by then. Otherwise, we will need to develop a way to apply the sensitivity score and then cut off any items beyond the max fraction of anomalies. This works because each technique will apply a score to each data point and we can order these scores in such a way that higher scores are more likely to be outliers. With these scores, even if the technique we use has no concept of sensitivity or a cutoff point, we can implement that ourselves.

Combined Arms

The more difficult scenario is when we need to create *ensemble models*, that is, models with several input algorithms. Here, we will need a bit more control over which items get marked as outliers, as each technique in the ensemble will have its own opinion of the outlier-worthiness of a given data point. We will need to agglomerate the results and perform the final scoring and cutoffs ourselves.

We will also want to provide weights for each input algorithm. Some techniques are more adept at finding outliers than others, and so we will want to weigh them more heavily. But there is a lot of value in incorporating a variety of input strategies, as no single technique is going to be perfect at finding all anomalies and ignoring all noise. For this reason, even relatively noisier techniques can still provide value, especially when there are several of them. Our hope is that the noise "cancels out" between the techniques, leaving us with more anomalies and (ideally!) less noise.

In the simplest case, we can weight each of the techniques equally, which is the same as saying that we let each technique vote once and count the number of votes to determine whether a particular data point is an outlier or not. With larger numbers of techniques, this may itself be a valid criterion for sensitivity score—suppose we have 20 separate tests. We could give each test a weight score of 5, giving us 5 * 20 = 100 as our highest score. This aligns quite nicely with sensitivity score, so if the caller sends in a sensitivity score of 75, that means a particular data point must be an outlier for at least 15 of the 20 tests in order to appear on our final list.

With differential weighting, the math is fundamentally the same, but it does get a little more complicated. Instead of assigning 5 points per triggered test, we might assign some tests at a value of 10 and others 2 based on perceived accuracy. The end result is that we still expect the total score to add up to 100, as that lets us keep the score in alignment with our sensitivity score.

Regardless of how we weight the techniques, we will want to send back information on how we perform this weighting. That way, we will be able to debug techniques and get a better feeling for whether our sensitivity score is working as expected.

Framing the Solution

With the context of the prior section in mind, let's revisit the code in Listing 4-3 and expand the solution to include everything we need. As far as inputs go, we will need to define `input_data` as the correct kind of list. FastAPI includes a concept called `BaseModel` that allows us easily to interpret the JSON our caller passes in and convert it into an object that Python understands. For univariate statistical input, we will need a single `value` column. It also would be proper to add a `key` column as well. That way, if the caller wishes to assign specific meaning to a particular value, they can use the key column to do so. We will not use the key as part of data analysis but will retain and return it to the user.

The detect_univariate_statistical() function should take in the sensitivity score and max fraction of anomalies and output the weights assigned for particular models, as well as any other details that might make sense to include for debugging.

Listing 4-4 shows an updated version of the univariate detection function with the addition of sensitivity score and the maximum fraction of anomalies as inputs, as well as weights and model details as outputs.

Listing 4-4. The newly updated API call for univariate outlier detection

```
class Univariate_Statistical_Input(BaseModel):
  key: str
  value: float

@app.post("/detect/univariate")
def post_univariate(
  input_data: List[Univariate_Statistical_Input],
  sensitivity_score: float = 50,
  max_fraction_anomalies: float = 1.0,
  debug: bool = False
):
  df = pd.DataFrame(i.__dict__ for i in input_data)

  (df, weights, details) = univariate.detect_univariate_statistical(df,
  sensitivity_score, max_fraction_anomalies)

  results = { "anomalies": json.loads(df.to_json(orient='records')) }

  if (debug):
    # TODO: add debug data
    # Weights, ensemble details, etc.
    results.update({ "debug_msg": "This is a logging message." })
    results.update({ "debug_weights": weights })
    results.update({ "debug_details": details })
  return results
```

The other methods will have very similar method signatures, although each of the calls will require its own model. For multivariate outlier detection, we will replace the single float called value with a list of values. Single-input time series anomaly detection will bring back the singular value but will include a dt parameter for the date and time. Finally, multi-series time series anomaly detection will add the date and also a series_ key value, which represents the specific time series to which a given data point belongs.

After sketching out what the API should look like—and understanding that we are not yet complete, as we will need to incorporate debugging information as we design the process—we can then stub out each of the four detection methods. A *stub* or *stub method* is a technique for rapid development in which we hard-code the output signature of a function so that we can work on the operations that call this function before filling in all of the details. Creating stub methods allows us to solve the higher-level problem of creating our API and ensuring the API code itself works before trying to tackle the much tougher job of implementing outlier detection. Then, when we are ready to dive into each technique, we already have a shell of the code ready for us to use.

To implement our stub methods, we will create a models folder and then one file per detection method. Listing 4-5 shows what the stub method looks like for detect_ univariate_statistical.

Listing 4-5. The stub method for detecting univariate outliers via the use of statistical methods. The output is hard-coded to return something that looks like a proper response but does not require us to implement the underlying code to generate correct results.

```
import pandas as pd

def detect_univariate_statistical(
  df,
  sensitivity_score,
  max_fraction_anomalies
):
  df_out = df.assign(is_anomaly=False, anomaly_score=0.0)
  return (df_out, [0,0,0], "No ensemble chosen.")
```

The other functions will look similar to this, except that there may be additional parameters based on the specific nature of the call—for example, multivariate outlier detection requires a parameter for the number of neighboring points we use for analysis. Once those are in place, we've done it: we have a functional API server and the contract we will provide to end users; if they pass in data in the right format and call the right

API endpoint, we will detect outliers in that given dataset and return them to the end user. The last thing we should do in this chapter is to package this up to make it easier to deploy.

Containerizing the Solution

We have a solution and we can run it as is on our local machines, but suppose we wish to deploy this solution out to a service like Azure Application Services. How would we package up this code and make it available to run? One of the easiest ways of packaging up a solution is to containerize it. The big idea behind containers is that we wish to pull together code and any project dependencies into one single *image*. This image can then be deployed as an independent *container* anywhere that runs the appropriate software. I can package up the software from this API and make the image available to you. Then, you can take that image and deploy it in your own environment without needing to install anything else.

Installing and configuring containerization solutions like Docker is well outside the scope of this book. If you have or can install the appropriate software, read on; if not, you can run everything in this book outside of containers with no loss in understanding.

For Windows users, there is, at the time of this writing, one major product available for working with containers: Docker Desktop. Docker Desktop is free for personal, noncommercial use. For Mac users, Docker Desktop works, but there is also a free alternative called Lima. Linux users have a wide variety of containerd-based options, although I am partial to Moby. Regardless of the product you use, the *Dockerfile* will be the same for use. Listing 4-6 shows the Dockerfile we will use to start up a containerized version of our anomaly detector.

Listing 4-6. The contents of the Dockerfile we will use in this book

```
FROM python:3.9
WORKDIR /code
COPY ./requirements.txt /code/requirements.txt
RUN pip install --no-cache-dir --upgrade -r /code/requirements.txt
COPY ./src/app /code/app
CMD ["uvicorn", "app.main:app", "--host", "0.0.0.0", "--port", "80"]
```

Python version 3.9 is the latest version as of the time of writing, so we will use that. The rest of this is fairly straightforward: install Python packages from the requirements. txt file, move our application to its resting place, and run uvicorn.

If you wish to use the containerized solution, you can find additional instructions in the readme for the GitHub repo associated with the book at `https://github.com/ Apress/finding-ghosts-in-your-data`.

Conclusion

In this chapter, we started breaking ground on our anomaly detection solution. That includes deciding on the "plumbing" in terms of protocols and base applications, as well as creating the interface that our callers will use as they input data and expect a list of outliers back. In the next chapter, we will create an accompanying test project to ensure that as we change code throughout the book, everything continues to work as expected.

CHAPTER 5

Building a Test Suite

Throughout the course of the prior chapter, we started to put together an application, stubbing out a series of useful methods. In this chapter, we will put together some tests, allowing us to ensure that the changes we make throughout the book will not break existing functionality or lead to undesirable results. First, we will look at a variety of tools available to us. Then, we will cover a few tips for writing testable Python code. Finally, we will create a set of unit tests and a set of integration tests, giving us the capability to run them at any time to ensure that we maintain product quality.

Tools of the Trade

There are two major varieties of test that we will cover in this chapter: *unit tests* and *integration tests*. These are not the only types of tests we could create, but they are two of the most important types.

The purpose of a unit test is to exercise some narrow capability of an application's code. Generally, we create unit tests against individual functions or methods in the code base, exercising that function with a variety of normal, abnormal, and even outlandish inputs to ensure that the code behaves appropriately in all of these circumstances. For example, we have a method called detect_univariate_statistical() for univariate outlier detection. This method takes in three inputs: a Pandas DataFrame, a numeric sensitivity score, and a numeric value representing the maximum fraction of anomalies allowed. It returns a Pandas DataFrame, a collection of weights, and a string representing additional details. With our tests, we will want to exercise different scenarios and see if we find incorrect results or a combination that results in an error.

Integration tests, meanwhile, are intended to exercise the combination of code components. Our unit tests determine if a given function works as expected given some set of inputs, and our integration tests ensure that the broader application behaves as

© Kevin Feasel 2022
K. Feasel, *Finding Ghosts in Your Data*, https://doi.org/10.1007/978-1-4842-8870-2_5

expected. Because our outlier detection application runs on FastAPI, the most obvious integration testing process would drive from those API calls, as they represent the way that end users will operate our system.

These high-level explanations of unit and integration testing apply regardless of the language. In the next section, we will narrow in on choosing libraries and tools for each purpose.

Unit Test Library

Python has a variety of unit testing libraries available, but we will use the pytest framework (`https://docs.pytest.org`). This is a low-ceremony library, meaning that you can easily integrate it in with an existing application and do not need to write a lot of code to create and run tests. Many large Python projects—such as scikit-learn—use pytest, so we will follow in their footsteps.

Pytest is easy to install on a machine by running `pip install pytest`. Once it is installed, we can begin writing tests. We will create all of our unit tests in the `\test\` directory, outside of our `\src\` source code directory. That way, we could later package the source code itself independent of these unit tests. It also creates a clean separation between application code and test code, allowing different developers to work on tests vs. code without the risk of interference.

When it comes to automating unit tests, the tox automation project (`https://pypi.org/project/tox`) allows us to run continuous integration tasks, including sets of unit tests, from the command line. It integrates quite well with pytest, which makes sense considering that some of the authors of tox are also pytest authors. We will not implement unit test automation in this book, but it would be a natural next step.

Integration Testing

There are plenty of integration testing applications and libraries available to us. One of my favorites for working with REST APIs is called Postman (`https://postman.com`). Postman includes a desktop application with support for Windows, macOS, and Linux. It is easy to create one-off tests in Postman, but you can also create collections of tests and run them from the Postman application, through the Postman service, or using a Node Package Manager (npm)–based collection runner as part of an automated solution.

Writing Testable Code

There are practices we can follow to write code in Python that is easier to test than the alternatives. Some of these practices are common across most general-purpose programming languages (such as C#, F#, Java, Scala, and the like), but others are particular to Python. In this section, we will look at several ideas which should make the development of tests a little easier in Python.

Keep Methods Separated

First, we want to limit the size and scope of Python methods (and functions). A good idea is to have a Python method do one thing very well. It may not always be possible to limit a method to one thing (and your definition of "one thing" might differ from mine in specific scenarios), but as a rule of thumb, break out as much as you can. This provides two specific benefits: one obvious and one less so. The obvious benefit is that we can test methods independent of one another, making it easier to isolate and test functionality. The less obvious benefit is that by breaking code out into component methods, we have the opportunity to test not only the specific use cases our current code covers but other logical use cases as well.

For example, suppose we have a process that calculates the payoff amount for a loan given the amount of the loan remaining, the interest rate of the loan, and the expected date of payment. If this calculation lives in a method that retrieves loan details for a customer, calculates the payoff amount for seven days from now, and does a check to see if the customer is eligible for any rate adjustments, testing this functionality can be quite difficult because we would need to build a matrix of possible customers: some who have outstanding loans and others who don't; for each of those two cases, a variety of loans at different interest rates; and also variants of each in which the customer is eligible for a rate adjustment. Even if we suppose that there are two classes of customer (has a loan, doesn't have a loan), six major cases for loan payout (which might be related to the amount remaining, the interest rate, whether the loan is fixed rate or adjustable rate, and if the customer is allowed to pay the loan off early without penalty), and three types of rate adjustment (not eligible, eligible for a small change, eligible for a large change), we have 2*6*3=36 separate test cases for this one function. That's a lot of testing for one method!

By breaking these out into three separate methods, we significantly reduce the number and complexity of tests: we need two tests to cover a method to retrieve loan details for a customer, six cases to cover various loan payout contingencies, and three cases to handle rate adjustment projections. Now, instead of 36 test cases, we have 2+6+3=11. Add in a couple of integration tests and we might have 13 tests, which is much easier to maintain than 36.

Furthermore, it's easier to think through and test edge cases with smaller, single-purpose methods. For example, if the nominal interest rate becomes negative, will this break our loan calculations? Several individual European central banks, the Bank of Japan, and the European Central Bank have all set negative nominal interest rate policies over the past 10–15 years; even so, it is unlikely that the developer who implemented this method thought about the possible ramifications of negative loan rates, but by keeping methods separated and single purpose, it becomes much easier to test this kind of special case.

Emphasize Use Cases

Following on from the first point, our code should work to implement specific business use cases, in which *actors*—that is, people or systems that interact with our application—have needs or desires and it is our application's job to fulfill those desires. For example, in our outlier detection algorithm, an end user wants to determine which points in a set of time series data points are anomalous. These specific use cases may require quite a few functions to get the job done, but we want to keep the use case in mind and develop tests that ensure that we can satisfy the use case. These tests will typically be integration tests rather than unit tests, as we want to ensure that the behavior of the application is such that we satisfy the actor's desires.

Functional or Clean: Your Choice

The last point is around choosing a path for software development. We can write high-quality, testable Python code in two different ways, taking either a functional approach or an object-oriented approach.

The functional approach emphasizes small, *deterministic* functions. By deterministic, I mean functions that, given a particular set of inputs, always return the same outputs. As an example, translating temperatures from Celsius to Fahrenheit (or vice versa) is a deterministic operation. If it is 9 degrees Celsius, it is 48.2 degrees Fahrenheit. No matter how many times we call the conversion function, it should always return the same value. As a counterexample, getting the logged-in user's remaining loan amount will be nondeterministic in nature. The loan amount itself is not strictly dependent on the inputs: if I send in the logged-in user ID 27, I am not guaranteed to get a value back of $14,906.25. I might get that value *this time*, but as soon as the user makes a loan payment and that number drops, the relationship no longer holds. Therefore, the next time I run the test, the result may or may not match my test's expectations, leading to spurious test failures. By making functions deterministic, our tests are less likely to break, and we will spend less time fixing failing tests.

Another important aspect of functional programming relevant to writing testable Python code is that functions should not have *side effects*. In other words, functions take inputs and convert them to outputs; they don't do anything else. This approach is aspirational rather than entirely realistic—after all, saving to the database is a side effect, and most applications would be fairly boring if they offered absolutely no way to modify the data. It just happens to be the case that our outlier detection engine can be close to side effect-free because we do not create files, save to a database, or push results to some third-party service. With most applications, however, we do not tend to be so lucky.

In practice, "no side effects" really means "have as many functions as possible be side effect-free." The biggest benefit to this is that it is easier to reason over your code: I can see a function that calculates monthly interest payments given a principal amount, an interest rate, and number of payments. If that is all the code does, a single method call like `monthly_payment = get_monthly_payment(principal, irate, num_payments)` is all we need to see. If the `get_monthly_payment()` function also assigns a loan representative to the customer, sends an email to the customer, and fires off a job to update interest rates in the database, we would need carefully to read through the contents of the function to understand what is happening. This would make testing considerably more difficult. By making as many functions as possible side effect-free, we simplify the process of test design and creation.

In addition to the functional programming approach, we can also take an object-oriented approach to Python development. Sebastian Buczyński applies the concept of the Clean Architecture, a take on object-oriented programming, specifically to Python in his blog at `https://breadcrumbscollector.tech/python-the-clean-architecture-in-2021/`. If the approach suits you well, Buczyński (2020) is a book-length treatment on the topic.

My personal biases push me toward functional programming, and so the code in this book will skew functional. That said, there is a lot of value in writing object-oriented Python and taking advantage of Clean Architecture for fully featured applications.

Creating the Initial Tests

Let's now create a few tests, giving us the basis for future application development. We will first start with a set of unit tests and then move on to some integration tests. Because all we have so far is a series of stub methods, the most we can test is that the shape of results matches our expectations. As we develop more code, we should be able to expand this test library and turn it into something more robust.

Unit Tests

Our first unit test will cover the simplest case: univariate outlier detection. Create a file named `test_univariate.py` in the \test\ folder. The pytest library will only look at files starting with the name "test" or ending with the name "test," so we will need that word in our file names.

Inside the file, pytest will only consider a method to be a test if it starts with "test" or is in a class whose name starts with "Test." We will start each relevant test method with "test_" to make finding tests easier. The first test method we will create is called `test_detect_univariate_statistical_returns_correct_number_of_rows`. Despite the name being quite the mouthful, it does tell us exactly what the test does: it calls `detect_univariate_statistical()` and makes sure that if we pass in a certain number of records, we get back that same number of records. The implication here is that we do not add or lose any data points from our dataset while performing outlier detection.

Listing 5-1 gives us a simple example of what this test method looks like.

Listing 5-1. A simple test method

```
from src.app.models.univariate import *
import pandas as pd
import pytest

def test_detect_univariate_statistical_returns_correct_number_of_rows():
  # Arrange
  df = pd.DataFrame([1, 2, 3, 4, 5, 6, 7, 8, 9, 10])
```

```
sensitivity_score = 75
max_fraction_anomalies = 0.20
# Act
(df_out, weights, details) = detect_univariate_statistical(df,
sensitivity_score, max_fraction_anomalies)
# Assert: the DataFrame is the same length
assert(df_out.shape[0] == df.shape[0])
```

The first line of code is how we get our reference to detect_univariate_
statistical(). Because we have an __init__.py file in \src\, pytest is able to pick
it up. After defining the test method, we have three sections: Arrange, Act, and Assert.
This nomenclature of unit testing gives us a consistent pattern for writing tests. In the
Arrange section, we perform all of the necessary setup. First, we create a sample Pandas
DataFrame containing ten elements, as well as the other input parameters we will need
for this test. Because the sensitivity score and max fraction of anomalies should not be
pertinent to the test results, we will set them at reasonable values for this first test.

After arranging our inputs, we Act upon them, calling the method under test. We
pass in the inputs and expect to get back a tuple of outputs, including another Pandas
DataFrame, an array containing weights, and a string with additional details. The Act
portion of a unit test is typically rather small, often just one line.

Finally, we have the Assert phase of a test. In the Assert phase, we perform the actual
testing using a series of assertion statements. With pytest, we do this with the assert()
function. Ideally, we perform one assertion, as that indicates that we are testing a single
thing. In practice, we might have multiple assertions to capture additional relevant
angles of the problem.

One of the nicest things about building unit test libraries is that it is easy to make
sure you are doing things correctly: you can run pytest in a terminal in the test directory
and get the results that you see in Figure 5-1.

```
PS F:\Book Development\Finding Ghosts in Your Data\code\test> pytest
============================================= test session starts =============
platform win32 -- Python 3.8.8, pytest-6.2.3, py-1.10.0, pluggy-0.13.1
rootdir: F:\Book Development\Finding Ghosts in Your Data\code\test
plugins: anyio-3.4.0
collected 1 item

test_univariate.py .

============================================= warnings summary =============
E:\Anaconda3\lib\site-packages\pyreadline\py3k_compat.py:8
  E:\Anaconda3\lib\site-packages\pyreadline\py3k_compat.py:8: DeprecationWarning: Using or importing the ABCs fror
Python 3.3, and in 3.9 it will stop working
    return isinstance(x, collections.Callable)

-- Docs: https://docs.pytest.org/en/stable/warnings.html
============================================= 1 passed, 1 warning in 0.55s =======
PS F:\Book Development\Finding Ghosts in Your Data\code\test> ▌
```

Figure 5-1. *A single test has run successfully*

The way that pytest marks test results is using a green period, which may be difficult
to see in print (or on a screen). Figure 5-2 shows an example of a test failure. Successful
tests tend to fade into the background, but failed tests—marked as a red F near the file
name—are very clear.

```
PS F:\Book Development\Finding Ghosts in Your Data\code\test> pytest
============================================= test session starts =============================================
platform win32 -- Python 3.8.8, pytest-6.2.3, py-1.10.0, pluggy-0.13.1
rootdir: F:\Book Development\Finding Ghosts in Your Data\code\test
plugins: anyio-3.4.0
collected 1 item

test_univariate.py F

================================================ FAILURES ================================
_____ test_detect_univariate_statistical_returns_correct_number_of_rows2 _

    def test_detect_univariate_statistical_returns_correct_number_of_rows2():
        # Arrange
        df = pd.DataFrame([1, 2, 3, 4, 5, 6, 7, 8, 9, 10])
        sensitivity_score = 0.75
        max_fraction_anomalies = 0.20
        # Act
        (df_out, weights, details) = detect_univariate_statistical(df, sensitivity_score, max_fraction_anomalies)
        # Assert:  the DataFrame is the same length
>       assert(df_out.shape[1] == df.shape[0])
E       assert 3 == 10

test_univariate.py:13: AssertionError
============================================= warnings summary =============================
E:\Anaconda3\lib\site-packages\pyreadline\py3k_compat.py:8
  E:\Anaconda3\lib\site-packages\pyreadline\py3k_compat.py:8: DeprecationWarning: Using or importing the ABCs from 'collection:
Python 3.3, and in 3.9 it will stop working
    return isinstance(x, collections.Callable)

-- Docs: https://docs.pytest.org/en/stable/warnings.html
============================================= short test summary info =======================
FAILED test_univariate.py::test_detect_univariate_statistical_returns_correct_number_of_rows2 - assert 3 == 10
============================================= 1 failed, 1 warning in 0.79s =====================
PS F:\Book Development\Finding Ghosts in Your Data\code\test> ▯
```

Figure 5-2. *An example of a failed test*

In this case, we can see the definition of the failed test, the error message, the line of code relevant to this failure, and a summary of the problem in the "short test summary info" section. With this, we can troubleshoot and correct the issue.

One test is great, but we will often wish to test several different sets of input parameters. For example, our test works great with a set of ten elements, but how would it work against other DataFrames? Will it work just as well if we have a single element in our DataFrame? What about an empty DataFrame? We could create separate tests for each of these, but it would involve quite a bit of copy and paste, and frankly, our test names are long enough as it is! Fortunately for us, pytest includes the ability to run the same test with multiple sets of inputs. This allows us to try a variety of situations, including normal inputs, edge cases, and situations that previously caused bugs in the code. In Listing 5-2, we use the `@pytest.mark.parameterize()` indicator to try out several possible tests. Note that for this to work, we also need to add a new input parameter for the test method: our input list for the Pandas DataFrame.

Listing 5-2. Using the parameterize method in pytest to run the same test with several input datasets

```
@pytest.mark.parametrize("df_input", [
  [1, 2, 3, 4, 5, 6, 7, 8, 9, 10],
  [1],
  [1, 2, 3, 4.5, 6.78, 9.10],
  []
])
def test_detect_univariate_statistical_returns_correct_number_of_
rows(df_input):
  # Arrange
  df = pd.DataFrame(df_input)
  sensitivity_score = 75
  max_fraction_anomalies = 0.20
  # Act
  (df_out, weights, details) = detect_univariate_statistical(df,
  sensitivity_score, max_fraction_anomalies)
  # Assert: the DataFrame is the same length
  assert(df_out.shape[0] == df.shape[0])
```

Running this gives us the results in Figure 5-3. Now you can see one green dot per successful test, or four in total.

```
PS F:\Book Development\Finding Ghosts in Your Data\code\test> pytest
============================================= test session starts =================
platform win32 -- Python 3.8.8, pytest-6.2.3, py-1.10.0, pluggy-0.13.1
rootdir: F:\Book Development\Finding Ghosts in Your Data\code\test
plugins: anyio-3.4.0
collected 4 items

test_univariate.py ....

============================================= warnings summary ===================
E:\Anaconda3\lib\site-packages\pyreadline\py3k_compat.py:8
  E:\Anaconda3\lib\site-packages\pyreadline\py3k_compat.py:8: DeprecationWarning: Using or importing the ABCs from '
Python 3.3, and in 3.9 it will stop working
    return isinstance(x, collections.Callable)

-- Docs: https://docs.pytest.org/en/stable/warnings.html
============================================= 4 passed, 1 warning in 1.55s ==========
PS F:\Book Development\Finding Ghosts in Your Data\code\test> []
```

Figure 5-3. *Multiple successful tests with no failures*

The test folder in the accompanying GitHub repository includes these tests as well as tests for each of the other techniques. This will give us enough of a foundation for unit tests, at least for now. One part of most subsequent chapters will be to expand the list of tests, whether because we add new functionality or simply to retain a case that our outlier detection engine failed to recognize. With this in place, let us turn our attention to the other side of testing: integration tests.

Integration Tests

Our unit tests will help ensure that the functionality in our outlier detector works as expected. Integration tests will help ensure that the experience a user receives is consistent with our expectations. The Postman application makes it easy to create API calls and also put together tests to ensure that the responses are what we expect. One big difference between unit tests and integration tests, at least in the context we are using those terms, is that unit tests can run independently. Pytest includes its own capability to run tests and call code separate from our FastAPI-based service, and so the service does not need to be running on a particular machine for our unit tests to work. By contrast, the point of integration tests is to check the behavior of a system of interconnected parts, and the easiest way for us to do this is to have the application running and call API endpoints. Thus, for the following tests to run, you will need to have uvicorn currently running and hosting the API project.

Figure 5-4 shows an example of a Postman API test, calling the univariate endpoint and passing in an array with two JSON-serialized Univariate_Statistical_Input objects. We also set the debug parameter to be False, meaning that we will not get debug information back from the API.

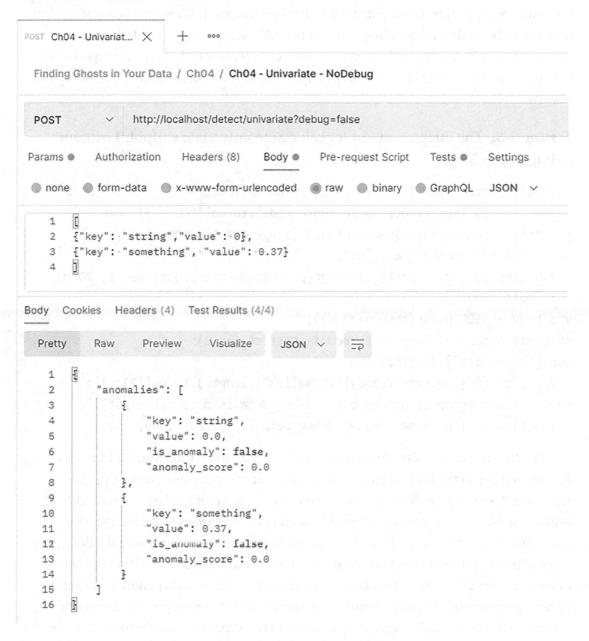

Figure 5-4. *A Postman test and its corresponding results*

What we see in Figure 5-4 is something we could easily replicate in other tools, including something as simple as submitting a series of web requests using the cURL application. What makes Postman interesting is its ability to write JavaScript-based tests against the response object and display the results in a Test Results section. Listing 5-3 contains four separate tests against the univariate endpoint. These tests look at separate aspects of the problem, including whether the call itself was successful, the correctness of specific results, and even the quality of service in terms of whether the response time is reasonable (though in this case, a response time of seven seconds may not always be reasonable!).

Listing 5-3. The integration test for calling the univariate endpoint without debugging enabled

```
pm.test("Status code is 200: " + pm.response.code,
    function () { pm.expect(pm.response.code).to.eql(200); });
pm.test("Response time is acceptable (range 5ms to 7000ms): " +
pm.response.responseTime + "ms",
    function () { pm.expect(_.inRange(pm.response.responseTime, 5, 7000)).
    to.eql(true); });
var body = JSON.parse(responseBody);
pm.test("Number of items returned is correct (eq 2): " +
body["anomalies"].length,
    function () { pm.expect(body["anomalies"].length).to.eql(2); });
pm.test("Debugging is not enabled (debug_details does not exist).",
    function () { pm.expect(body["debug_details"]).undefined; });
```

The first test checks that the response code is 200, which indicates that the service functioned correctly. We passed in a reasonable set of inputs and expect a proper response. Then, we check the response time and ensure that it is between 5 and 7000 milliseconds. I like to have a minimum boundary here to protect against a service immediately (and incorrectly) returning a success response without actually doing any work. Finding a proper lower bound may require a few iterations, but if success never takes less than 50 or 100 milliseconds, a success in 3 milliseconds generally indicates either that your code became significantly faster—in which case you'll want to update the test—or something is happening to cause an improper conclusion to the test, for example, a bug that prevents running the code and simply returns an empty set.

The remaining tests look at specific elements in the response, and to get these elements, we will need to parse the response body. The third test ensures that we have two items returned. This matches the number of items we sent in, so we have neither lost nor gained data elements. The fourth test checks that we do not include any debugging details, as we set the debug flag to False. Figure 5-5 includes the tests and the results. We can see clear textual indicators for each test, making it easy to compare expectations vs. actual results for each test associated with our particular request.

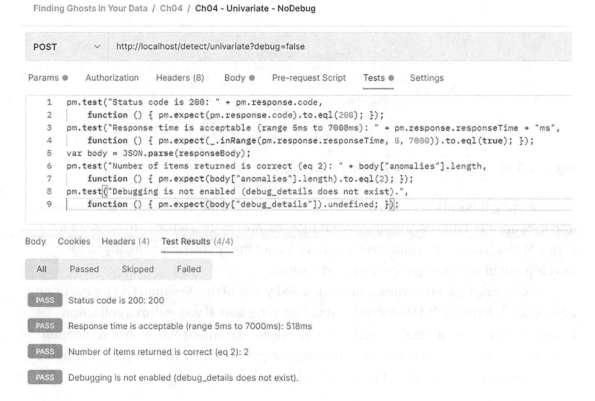

Figure 5-5. *The test results for a single Postman API request*

This is certainly useful in itself, as we can save this request and run it again in the future to ensure that fundamental behavior has not changed. If we have a large number of API requests, we can bundle them together into a Postman collection. A Postman collection is a logical grouping of API requests, and within a collection, we can further break things down into folders. That's helpful for cataloging the requests, but Figure 5-6 shows the real utility of this: running all tests within a collection.

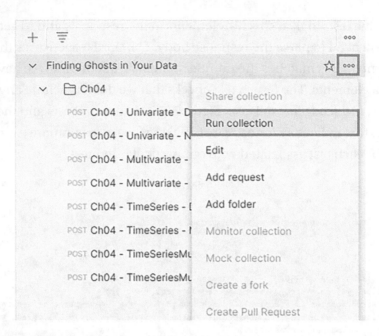

Figure 5-6. *Run all tests in a given collection*

Selecting this option brings us to a test runner, which allows us to run each request and execute all of the tests. Similar to our unit test library, this lets us put together a series of integration tests and ensure that we do not break core functionality as we develop out new techniques for outlier detection.

Automating collection runs is also possible by way of the Newman CLI for Postman, accessible at `https://github.com/postmanlabs/newman`. If you export a collection from Postman, you can make it available to a machine running Newman and schedule collection runs. This is not as easy as running the pytest command, but you can build a robust testing platform around Postman collections.

Conclusion

Throughout the course of this chapter, we started to create a set of unit tests using pytest and integration tests in Postman. We will continue to expand on these tests throughout the book. In the next chapter, we will return back to the outlier detection application and make our first attempt at solving univariate outlier detection using statistical techniques.

CHAPTER 6

Implementing the First Methods

The prior two chapters have given us a foundation for work, with Chapter 4 laying out the fundamental API calls and stub methods and then Chapter 5 following up with unit and integration tests. In this chapter, we will build upon that foundation and begin to implement our first outlier detection tests. Following the principle of "start with simple," we will put into place some of the easiest tests: statistics-based, univariate outlier detection tests.

A Motivating Example

A univariate outlier detector sounds rather limiting: almost no phenomenon is so simple that we can boil it down to a single variable. In practice, though, univariate anomaly detection can be extremely powerful, as our purpose is not to explain the world but rather to understand if something strange has happened. In this regard, as long as we have some measure that is close enough to being independent from other measures, we can perform a univariate outlier detection analysis and get our answers.

For example, your PC is a complicated mixture of parts, and it would be folly to believe that any single measure could completely describe the health of your machine. There are, however, single measures that can describe the health of some part of your PC. For example, CPU temperature is a great measure of how well your computer's cooling systems are working, and there are optimal ranges based on ambient temperature and workload. A quick rule of thumb is that under heavy load, a modern CPU will typically run anywhere from 60 to 80 degrees Celsius (140–176 degrees Fahrenheit), but temperatures above this can be dangerous. If we track the output of a CPU's thermal sensor by collecting the average temperature over common intervals (e.g., ten-second or one-minute intervals), we could use univariate statistical analysis

© Kevin Feasel 2022
K. Feasel, *Finding Ghosts in Your Data*, https://doi.org/10.1007/978-1-4842-8870-2_6

to check if there are temperatures well above the norm. These high temperatures could indicate a problem with some aspect of our cooling solution, extreme load on the system, or a problem with the thermal sensor. The goal here is not to determine the cause but to provide information to an operator that something deserves the operator's attention.

In the preceding example, we focused on extremely high temperatures, as that is a common problem we experience with PCs. For portable devices, especially devices that survive outdoors, we need to be concerned not only with high temperatures but also with low temperatures, as extremely low temperatures might also damage electronic components or indicate an issue with a thermal sensor. Our tests in this chapter will focus on *bidirectional* outlier detection, meaning we will alert on abnormally high values as well as abnormally low values. It would be possible to include unidirectional analysis as an option, but that is an exercise left to the reader.

Now that we know this outlier detection will span in both directions, we will want to determine which measure—or measures—we should use. In Chapter 3, we looked at a variety of statistical measures, such as number of standard deviations from the mean, distance from the interquartile range, and number of median absolute deviations from the median. Given these tests, as well as all of the other statistical tests we could look at, which one of these should we use? In this case, we will embrace the power of "and."

Ensembling As a Technique

The tests we will use in this chapter are all fairly simple statistical tests. The major benefit to these tests being simple is that they are also quite fast. The major drawback to these tests being simple is that we do not have much opportunity to tune these tests and make them less noisy. This is where a technique known as *ensembling* can be quite beneficial to us.

Ensembling is the process of combining together a series of models and deriving a single answer from the combination of all models. The intuition behind this technique is fairly straightforward. There exists some set of points that make up our collection of data. We, as humans, may look at the data, but we cannot look at it until after our outlier detection engine has finished. Therefore, our outlier detection process cannot know beforehand which points are outliers, and so even though we have a set of all possible techniques we can use, we might not know when particular techniques are reasonable in this case. In other words, each technique is sometimes effective and sometimes ineffective.

The real key here is that the range of effectiveness for different techniques will differ—that is, Test A might fail under one set of circumstances, Test B might fail under another, and Test C under a third. Ideally, these three sets of failure conditions would not overlap at all, meaning that at least two out of the three tests will be correct for each data point and we simply choose the answer with the most votes to determine whether something is an outlier. Outlier detection in practice will not be this simple, although the intuition is not completely wrong: despite there being ranges of "failure region" overlap between techniques, different types of techniques can be more effective within certain ranges of data and the overlap is nowhere near total. For this reason, creating an ensemble of moderately effective tests can lead to a result that is superior to any single one of the tests.

There are two types of ensembling that we will discuss in this chapter: sequential and independent ensembling. Let's look at each one in turn.

Sequential Ensembling

The first type of ensembling technique we will look at is called *sequential ensembling*. The idea behind sequential ensembling is that we run each test in turn and the results from the prior test feed the subsequent test. In this case, we would first run Test A. The information on whether data points are outliers or inliers would then be fed alongside the raw data into Test B. Test B takes all of these inputs and then creates its own set of outliers and inliers. We take those and feed them into Test C and so on through all of our tests. The end result is a single set of data points with some consensus marking of whether a point is an outlier or an inlier.

Generally, sequential ensembling works by narrowing down the relevant set of potentially controversial data points. If the first model is extremely confident in a prediction, we might not want subsequent models to waste time on those data points. Instead, subsequent models try to answer the questions the first model found difficult. For this reason, we usually want the best model to go first and take care of as many data points as it can.

When using sequential ensembling, there is no guarantee that we will find a model that is supremely confident (for whatever definition of "supremely confident" we choose) for any given data point. Thus, by the end, we might have some data points none of the models conclusively define as either an outlier or an inlier. In that case, we could report back the results from the final test in the ensemble. Alternatively, we might break down to simple voting: with five tests, if three report that a data point is an outlier and two do not, we call it an outlier. These guesses will be fairly low confidence, but with a good sequence of tests, the aim is to have relatively few of these data points; instead, our goal is that the various tests in the ensemble will confidently predict answers for the vast majority of data points.

Independent Ensembling

Sequential ensembling can be great, but it does come with a risk: What if a model is confidently wrong? If our first model is very confident in a result but it turns out to be the wrong guess, none of the subsequent models will test it out and find this discrepancy. This is where *independent ensembling* comes into play. With independent ensembling, we still have a battery of tests that we perform on a dataset. The primary difference, however, is that these tests are done independently of one another—that is, the results of Test A are not fed into Test B, Test C, or any other test. Instead, each test is run over the entirety of the input data and generates predictions for every data point. Once we have all of the tests computed for all of the data points, we then need to combine together the results.

The simplest technique for combining together results is to count the number of times a test calls a given data point an outlier and see if we meet a certain threshold. Suppose we have five tests and require a simple majority to decide to label a given point an outlier. In that case, we would need three of the five tests to agree before we make the call, regardless of how confident any given test is in its prediction.

Another technique we can use is to weight the individual tests. For example, suppose that Test A tends to be fairly accurate and Test B is a bit less accurate on the whole but neither test is sufficiently accurate to stand on its own. In this case, we might give Test A more voting shares than Test B, such that if there is disagreement, Test A will override Test B. Adding in more tests, you can see how this fits together. Suppose we have Tests A, B, C, and D with weighted voting shares of 35, 15, 30, and 20, respectively. Furthermore, we declare that we need at least 50 shares for a given data point to be considered anomalous. In this case, there are several paths that can get us to 50 shares: A and anything else will give us at least 50 shares, for example. If only tests B and C declare something as an outlier, though, we will be short of our 50-share threshold, and therefore we will not mark the data point as an outlier.

Choosing Between Sequential and Independent Ensembling

Sequential and independent ensembling are both useful techniques for combining the results of multiple models or tests. Let's cover a few scenarios and understand when we might want to use one ensembling technique over the other.

First, if you have one particularly good algorithm, sequential ensembling can be superior to independent ensembling. This would allow us to rule out a large number of data points after the first test, making subsequent tests significantly faster. By contrast, independent ensembling would require that we run all data points against all tests, meaning that performance (in terms of speed) will not be as good.

By contrast, if you have a series of similar-quality algorithms but nothing that dominates, the independent ensembling technique is the safer approach. This will reduce the likelihood of incorrectly labeling an outlier as an inlier (or vice versa), and that can make up for independent ensemble being slower to run.

Second, if you know the performance characteristics of specific algorithms, a sequential ensembling process can make more sense. For example, under specific conditions, one algorithm might be more sensitive to outliers than another, but suppose that it works better on small numbers of rows than large numbers. In that case, you could try to gauge some of the characteristics of your dataset, determine if it more closely fits the first or the second algorithm, and lead with that one, using the other to act as a cleanup test. If you do not know exactly when one algorithm beats another, the independent ensemble is a safer choice.

Finally, if you have complementary checks, sequential ensembling will typically be better—that is, if there is relatively little overlap in the failure zones of each algorithm, feeding results sequentially and paring down the datasets will typically work well.

Implementing the First Checks

In this chapter, we will implement three algorithms: standard deviations from the mean, distance from the interquartile range, and median absolute deviations from the median. Reviewing our checklist from before, we do not have one algorithm that is head and shoulders above the rest. Median absolute deviations from the median is a more robust version of standard deviations from the mean, so the former will be better. That said, both will catch a lot, and there may be cases in which we want the added information from our test of standard deviations from the mean. Distance from the interquartile range covers a different space from our variance-based measures, but "different" is not guaranteed to be better here.

We also will not know the performance characteristics of these three algorithms. We can make some assumptions based on how they operate, but we haven't tried to define which algorithms work best with any given dataset.

Based on these two considerations, we will use independent ensembling in this chapter. Later on in the book, we will see an example in which sequential ensembling makes a lot of sense. In the meantime, let's write some code!

The first thing we will need to do is update the univariate.py file and add in our tests. Inside the detect_univariate_statistical() function, we will need to do four things: run our tests, score the results of those tests, determine which data points are outliers, and return the set of outliers. In order to make this code a bit more functional-friendly, we will perform each of these actions as independent functions, and each action will operate without side effects.

The first function, run_tests(), will be responsible for running each of the following tests.

Standard Deviations from the Mean

The first test we will implement checks whether a particular data point is more than a given number of standard deviations from the mean. We can call this function check_sd(), and it will run once for each data point in our dataset. The check_sd() function will take four parameters: the value (value), the mean of the dataset (mean), the standard deviation of the dataset (sd), and the minimum number of standard deviations' difference before we report an outlier (min_num_sd). The code for check_sd() makes up Listing 6-1.

Listing 6-1. An initial try at creating a standard deviations check

```
def check_sd(val, mean, sd, min_num_sd):
  if (abs(val - mean) < (min_num_sd * sd)):
    return abs(val - mean)/(min_num_sd * sd)
  else:
    return 1.0
```

This function checks to see if the absolute difference between our value and the mean is more than the minimum number of standard deviations. If the difference between our data point's value and the mean is within the acceptable range, we return the percentage of the way to our threshold of min_num_sd standard deviations. If the difference is outside of the acceptable range, we return a score of 1.0, indicating that this test considers the data point an outlier.

When we call check_sd(), we will know the data point's value, as the caller sends us this information. We will also know the minimum number of standard deviations, as we set this to 3. We will need to calculate the mean and standard deviation in the run_tests() function, something that is easy to do with Pandas DataFrames: df['value'].mean() and df['value'].std(), respectively.

Median Absolute Deviations from the Median

The second test we will implement in this chapter checks whether a particular point is more than a given number of median absolute deviations from the median. This is a robust version of our check for standard deviations from the mean, which means that a few outliers will not substantively affect the quality of our test. Because the code for this test is fundamentally similar to the prior test, we should create a common function for midpoint-based statistical tests. Listing 6-2 shows that function.

Listing 6-2. A new function serves to perform midpoint distance checks

```
def check_stat(val, midpoint, distance, n):
  if (abs(val - midpoint) < (n * distance)):
    return abs(val - midpoint)/(n * distance)
  else:
    return 1.0
```

With this in place, we can refactor the check_sd()function to call this new function. Listing 6-3 shows this as well as the new check_mad() function.

Listing 6-3. Three simple tests of midpoint distance

```
def check_sd(val, mean, sd, min_num_sd):
  return check_stat(val, mean, sd, min_num_sd)

def check_mad(val, median, mad, min_num_mad):
  return check_stat(val, median, mad, min_num_mad)
```

Distance from the Interquartile Range

The third and final test we will implement checks whether a particular point is more than a given number of interquartile ranges beyond the 25th and 75th percentiles. As a reminder, the interquartile range is the difference between the 75th and 25th percentiles of a dataset. We consider a value an outlier if one of two things is true: either it is more than 1.5 interquartile ranges above the 75th percentile or it is more than 1.5 interquartile ranges below the 25th percentile. The check_iqr() function will take six values: the value (value), the median of the dataset (median), the 25th percentile (p25), the 75th percentile (p75), the interquartile range of the dataset (iqr), and the minimum number of interquartile ranges from the relevant percentile before we report an outlier (min_iqr_diff). The code for check_iqr() makes up Listing 6-4.

Listing 6-4. An initial try at creating an interquartile ranges check

```
def check_iqr(val, median, p25, p75, iqr, min_iqr_diff):
  if (val < median):
    if (val > p25):
      return 0.0
    elif (p25 - val) < (min_iqr_diff * iqr):
      return abs(p25 - val)/(min_iqr_diff * iqr)
    else:
      return 1.0
  else:
    if (val < p75):
      return 0.0
    elif (val - p75) < (min_iqr_diff * iqr):
      return abs(val - p75)/(min_iqr_diff * iqr)
    else:
      return 1.0
```

This code differs a little bit from our first two functions, as we have two additional parameters. Instead of checking distance from the midpoint, we want to check against the 25th and 75th percentiles, but we only want to test each one in a single direction. Therefore, if our value is less than the median, we check against the 25th percentile. If the data point is between the 25th percentile and the median, our anomaly score for this test is 0—it is definitely an inlier. If we are beyond the 25th percentile but less than `min_iqr_diff` times the interquartile range below the 25th percentile, we calculate how close we are to the edge and generate a number in the range (0, 1). Any points beyond the cutoff will have an anomaly score of 1.0. For values greater than or equal to the median, we perform a similar test against the 75th percentile.

Now that we have these tests in place, we can move up one level and complete the `run_tests()` function.

Completing the run_tests( ) Function

We know that we want to have the `run_tests()` function call each test for each data point. We also know that this function needs to calculate aggregate measures, specifically the mean, standard deviation, median, interquartile range, and MAD. In order to do this, we need a Pandas DataFrame that contains a column called `value`. Listing 6-5 shows the implementation of this function.

Listing 6-5. The run_tests() function in its entirety

```
def run_tests(df):
  mean = df['value'].mean()
  sd = df['value'].std()
  p25 = np.quantile(df['value'], 0.25)
  p75 = np.quantile(df['value'], 0.75)
  iqr = p75 - p25
  median = df['value'].median()
  mad = robust.mad(df['value'])

  calculations = { "mean": mean, "sd": sd, "p25": p25, "median": median,
  "p75": p75, "iqr": iqr, "mad": mad }

  # for each test, execute and add a new score
  df['sds'] = [check_sd(val, mean, sd, 3.0) for val in df['value']]
  df['mads'] = [check_mad(val, median, mad, 3.0) for val in df['value']]
  df['iqrs'] = [check_iqr(val, median, p25, p75, iqr, 1.5) for val in
  df['value']]

  return (df, calculations)
```

This function accepts a DataFrame and then calculates the five relevant aggregate measures. Note that we use the `quantile()` function in numpy to calculate the 75th and 25th percentiles in our dataset. In addition, we use the `robust.mad()` function in the statsmodels package in Python to calculate MAD, as the `mad()` method in Pandas actually calculates the *mean* absolute deviation, not the *median* absolute deviation, and that minor difference in terms makes a huge difference in practice. The rest of the aggregation functions are methods available on a DataFrame.

After calculating the aggregates, we stick the results into a dictionary called `calculations`, which we will return back to the caller for debugging purposes. Then, we perform each check for each value in the DataFrame and add the results as a new column in our DataFrame. The ordering of data in the DataFrame will be the same for each list we create, so we do not need to sort the data before appending the new columns. Once we have completed this work, we return the updated DataFrame for the next step in the operation: scoring these results.

Building a Scoreboard

So far, we have a series of tests, each of which provides its own calculation of how anomalous a particular data point is. We now want to combine these independent tests and return back to the end user a unified anomaly score and classification of whether a given data point is anomalous.

When it comes to creating a scoring function, the check methods already give us a head start, as they provide two things. First, they give us a clear indicator of whether a test marks something as an outlier or an inlier. These delineations (3 standard deviations from the mean, 1.5 interquartile ranges from the 25th/75th percentile, 3 MAD from the median) may not be perfect, but they should serve us well at this point. Second, if we do find a data point that lies outside these margins, we get a direct indicator that we consider something to be an outlier. Including a threshold like this brings one benefit and one risk. The benefit is that an extreme outlier will not obscure less-extreme outliers—something 50 interquartile ranges from the 75th percentile is still an obvious outlier, even if another data point is 500 interquartile ranges away. The downside risk here comes from the fact that our capping of what constitutes an outlier might cause problems if the distribution of data is nothing like the normal distribution. In those cases (especially if we think back to the Cauchy distribution), we might end up with too many outliers, leading to more noise.

Weighting Results

Listing 6-6 provides code for a function to generate an anomaly score for each data point, based on the weighted results of each test.

Listing 6-6. Generate an anomaly score for each data point

```
def score_results(
  df,
  weights
):
  return df.assign(anomaly_score=(
    df['sds'] * weights['sds'] +
    df['iqrs'] * weights['iqrs'] +
    df['mads'] * weights['mads']
  ))
```

This function takes in a Pandas DataFrame as well as a dictionary of weights. If we wish to guarantee that anomaly score ranges between 0 and 1, the sum of all weights should add up to 1.0. Other than that potential linear constraint, we are free to choose whatever weights we desire.

Figuring out the correct weights will be a challenge, one that we should not pass off to our users. As a general principle, users should not need to know details about how we run our tests, nor should they need detailed information on the distribution of their data—after all, if they know all of this, why do they need our service? Therefore, we are going to need to come up with the weights directly. The easiest method for figuring out weights combines some basic intuition with a lot of trial and error. Let's start with the basic intuition.

When we have both robust and nonrobust statistics in our test set, we should bias toward the robust statistics, as they are more likely to remain sensitive in the face of some percentage of outliers. It would still be worth having both, as we can find scenarios in which a data point is more than 3 MAD from the median but not quite 3 standard deviations from the mean, so including both values will reduce our expectations around whether this particular data point is an outlier compared to a point that passes both tests. Still, biasing toward robust statistics would mean that we want to weight the MAD result higher than the standard deviation result. The number of interquartile ranges from the 25th or 75th percentile is also a fairly robust statistic and is a reasonably safe measure for tracking whether something is an outlier, so we will weight it greater than standard deviations from the mean but less than median absolute deviations from the median.

Coming up with exact numbers for these weights would require running statistical tests against a variety of relevant datasets for your particular environment and trying to ascertain which weights do the best job of finding outliers without being wrong too often. The biggest challenge here is that we typically do not have labeled data that indicates when there really was an anomaly; without that critical piece of information, we can track outliers but may not be able to guarantee that any given set of weights will be optimal for some environment.

For the purposes of moving forward, we will start with weights of 0.45, 0.35, and 0.25, respectively, for MADs from the median, IQRs from the median, and SDs from the mean. Those values do sum to a number greater than 1, which means that our anomaly score can be above 1. That's okay, though: if the values summed exactly to one, it becomes more difficult to find more than one anomaly in a given dataset without setting the sensitivity to a very low value like 20–30. In practice, even with these scores, we typically need at least two of the three tests to max out if we set the sensitivity score threshold to something like 75 or 80. I'd recommend keeping the sum of weights somewhere between 1.0 and 1.5, depending on whether you'd like to calibrate your system to show more outliers or fewer for a given sensitivity score.

Now that we have weights and can calculate an anomaly score, we can move to the final part: determining whether something is an outlier.

Determining Outliers

The reason we want to have an anomaly score that typically sits somewhere between 0 and 1 inclusive (but might go a little higher) is that the sensitivity score itself ranges from 1 to 100 inclusive. We can reverse and scale the sensitivity score to overlap with our anomaly score. Because sensitivity score ranges between 1 and 100 and 100 is the most sensitive, we will want to see more outliers with the higher scores. To do that, we can subtract the sensitivity score from 100, divide the result by 100.0, and compare that to our calculated anomaly score, allowing us to call something an anomaly if its anomaly score is greater than the sensitivity score cutoff. Well, almost. There is one little exception, and that has to do with the maximum fraction of anomalies.

In the event that our maximum fraction of anomalies is smaller than the fraction of data points whose anomaly score matches or beats the sensitivity score, we need to select the most anomalous data points up to that maximum fraction of anomalies. For example, suppose that 20% of our data points have anomaly scores greater than the passed-in sensitivity score but the user also indicated that the maximum fraction of anomalies should be 15%. In that case, we can only mark 15% of our data points as anomalous, and so we should choose the *most* anomalous results from the dataset. Listing 6-7 shows how to do this.

Listing 6-7. Determine whether a given value is an outlier, incorporating max fraction of anomalies

```
def determine_outliers(
  df,
  sensitivity_score,
  max_fraction_anomalies
):
  sensitivity_score = (100 - sensitivity_score) / 100.0
  max_fraction_anomaly_score = np.quantile(df['anomaly_score'], 1.0 - max_
  fraction_anomalies)
  if max_fraction_anomaly_score > sensitivity_score and max_fraction_
  anomalies < 1.0:
    sensitivity_score = max_fraction_anomaly_score
    return df.assign(is_anomaly=(df['anomaly_score'] > sensitivity_score))
```

The first step is to find the score at the `100 - (max_fraction_anomalies * 100)` percentile. That is, if `max_fraction_anomalies` is 0.1, we want to find the $(100 - 10) =$ 90th percentile of anomaly scores. If that score is less than the sensitivity score, it means that the maximum fraction of anomalies will not constrain our results and so we can use the sensitivity score. If that score is greater than the sensitivity score and the maximum fraction of anomalies is less than 1.0, it means that we should only return values greater than or equal to the max fraction's score. Regardless of which score wins out, we perform the determination of whether something is an anomaly the same way and return this as a new column.

This leaves Listing 6-8 to put all of the pieces together.

Listing 6-8. The new function to detect univariate outliers based on statistical tests

```
def detect_univariate_statistical(
  df,
```

```
  sensitivity_score,
  max_fraction_anomalies
):
  weights = {"sds": 0.25, "iqrs": 0.35, "mads": 0.45}

  if (df['value'].count() < 3):
    return (df.assign(is_anomaly=False, anomaly_score=0.0), weights, "Must
    have a minimum of at least three data points for anomaly detection.")
  elif (max_fraction_anomalies <= 0.0 or max_fraction_anomalies > 1.0):
    return (df.assign(is_anomaly=False, anomaly_score=0.0), weights, "Must
    have a valid max fraction of anomalies, 0 < x <= 1.0.")
  elif (sensitivity_score <= 0 or sensitivity_score > 100 ):
    return (df.assign(is_anomaly=False, anomaly_score=0.0), weights, "Must
    have a valid sensitivity score, 0 < x <= 100.")
  else:
    df_tested = run_tests(df)
    df_scored = score_results(df_tested, weights)
    df_out = determine_outliers(df_scored, sensitivity_score, max_fraction_
    anomalies)
    return (df_out, weights, { "message": "Ensemble of [mean +/- 3*SD,
    median +/- 1.5*IQR, median +/- 3*MAD].", "calculations": calculations})
```

In this function, we first set the weights. Then, we ensure that the caller passes in at least three data points, as we need a minimum of three data points to determine if something is an outlier. Two data points will never be enough to show an outlier, and fewer than two points will cause errors in calculating variance. Then, we check to ensure that the maximum fraction of anomalies and sensitivity score are set to reasonable values.

Assuming we have acceptable inputs, we iterate through each function call one at a time, starting with running tests, moving on to scoring results, and finally determining outliers. We end this by returning the output DataFrame, our list of weights, and a descriptor of what we tested. The descriptor includes the midpoint and distance calculations as well, making it easier to diagnose why certain data points have a given anomaly status.

Congratulations: you now have an outlier detector!

Updating Tests

Now that we have the code in place for outlier detection of univariate data via statistical tests, we can update the unit and integration test projects and ensure that the code works as we expect. First, we will update unit tests. Then, we will handle the integration tests.

Updating Unit Tests

In the prior chapter, our singular unit test provided us confidence that the changes we made would not fundamentally break the application. Incidentally, during development of the code in this chapter, this single unit test with a few cases did its job so well that it caught two separate bugs that might otherwise have snuck through. With that result making a unit test project already a net positive, let's add a few more unit tests now that we have more things to test.

There are three things we can effectively test at this state: the number of anomalies a particular dataset creates, how the sensitivity score affects the number of anomalies returned, and how the maximum fraction of anomalies affects the number of anomalies returned. Listing 6-9 includes the first new test, which includes cases in which we have a single anomalous value.

Listing 6-9. Test cases in which we expect to receive a single anomalous value

```
@pytest.mark.parametrize("df_input", [
  [1, 2, 3, 4, 5, 6, 7, 8, 9, 10, 90],
  [1, 1, 1, 2, 2, 3, 3, 4, 4, 5, 5, 5, -13],
  [0.01, 0.03, 0.05, 0.02, 0.01, 0.03, 0.40],
  [1000, 1500, 1230, 13, 1780, 1629, 1450, 1106],
  [1, 2, 3, 4, 5, 6, 7, 8, 9, 10, 19.4]
])
def test_detect_univariate_statistical_returns_single_anomaly(df_input):
  # Arrange
  df = pd.DataFrame(df_input, columns={"value"})
```

```
sensitivity_score = 50
max_fraction_anomalies = 0.50
# Act
(df_out, weights, details) = detect_univariate_statistical(df,
sensitivity_score, max_fraction_anomalies)
num_anomalies = df_out[df_out['is_anomaly'] == True].shape[0]
# Assert: we have exactly one anomaly
assert(num_anomalies == 1)
```

The test code is fairly straightforward: we arrange the input datasets as a Pandas DataFrame and set the sensitivity score to a static value of 50. We also set the maximum fraction of anomalies high enough that we could potentially trigger more than one anomalous result. Then, after calling the `detect_univariate_statistical()` function, we count the number of records in which `is_anomaly` is flagged as True and ensure that count is equal to 1.

Each of the test cases for this test serves a specific purpose. The first case shows a large increase in whole numbers. The second case incorporates a negative outlier, ensuring that we can handle negative values correctly. The third case looks at relatively small numbers, which could suss out a "minimum distance" bug. The fourth case ensures that we can find anomalous values that are smaller but not negative. Finally, the fifth case includes both whole numbers and decimal numbers, ensuring we can support both in the same call.

We can also create tests for other numbers of anomalies. In fact, we could create one large test that includes the dataset and the expected number of anomalies, though I would probably want to keep the cases of 0, 1, and 2 separate and include several test cases for each. Review the `test_univariate.py` file for these cases, although I do wish to focus down on one particular test case and my strategy behind it.

This particular test case has an input dataset of [1000, 1500, 2230, 13, 1780, 1629, 3202, 3025, 6]. A human looking at this might immediately spot 13 and 6 as outliers, but this is a case in which a human could be wrong. The most likely reason that a human would see 13 and 6 as outliers is the order of magnitude difference: 13 and 6 are two and three orders of magnitude smaller than the other values, respectively, so our eyes catch the difference. If we do the math, however, we'll see that they aren't quite as big of outliers as we might expect. The mean of this input dataset is 1598.33, and the median is 1629. When looking at these two measures, we can see that the values 3202 and 3025 are just as distant from the midpoint as 6 and 13. The 25th percentile of this dataset is 1000 and the 75th percentile is 2230, so the interquartile range is 1230 and 1.5 times the interquartile range is 1845. Finally, the standard deviation of this sample is 1143, and the MAD is 932.55. With these numbers, we can manually calculate all of our distances or we could simply call the outlier detection engine and get the weights. Figure 6-1 shows these results for 3202, 3025, and 6.

```
{
        "key": "7",
        "value": 3202.0,
        "sds": 0.467566763,
        "mads": 0.562253512,
        "iqrs": 0.5268292683,
        "anomaly_score": 0.554296015,
        "is_anomaly": false
},
{
        "key": "8",
        "value": 3025.0,
        "sds": 0.4159604543,
        "mads": 0.4989865879,
        "iqrs": 0.4308943089,
        "anomaly_score": 0.4793470862,
        "is_anomaly": false
},
{
        "key": "9",
        "value": 6.0,
        "sds": 0.4642624042,
        "mads": 0.5801255244,
        "iqrs": 0.5387533875,
        "anomaly_score": 0.5656857727,
        "is_anomaly": false
}
```

Figure 6-1. *Weights and anomaly scores for three non-outliers*

After doing the math, it's pretty obvious that, due to the spread in this dataset, 6 and 13 are not outliers, at least for a sensitivity score like 30. As the sensitivity score moves up, 6 and 13 become outliers but so do 3202 and 3025.

The next test we will review is to ensure that the sensitivity score correctly affects the number of anomalies we see in our results. Listing 6-10 shows the code for this test, as well as all of the test cases we will use.

Listing 6-10. Ensure that sensitivity score operates as expected

```
anomalous_sample = [1, 1, 2, 2, 3, 3, 4, 4, 5, 5, 6, 7, 8, 9, 10,
2550, 9000]
@pytest.mark.parametrize("df_input, sensitivity_score, number_of_
anomalies", [
  (anomalous_sample, 100, 17),
  (anomalous_sample, 95, 15),
  (anomalous_sample, 85, 8),
  (anomalous_sample, 75, 5),
  (anomalous_sample, 50, 2),
  (anomalous_sample, 25, 2),
  (anomalous_sample, 1, 1)
])
def test_detect_univariate_statistical_sensitivity_affects_anomaly_
count(df_input, sensitivity_score, number_of_anomalies):
  # Arrange
  df = pd.DataFrame(df_input, columns={"value"})
  max_fraction_anomalies = 1.0
  # Act
  (df_out, weights, details) = detect_univariate_statistical(df,
  sensitivity_score, max_fraction_anomalies)
  num_anomalies = df_out[df_out['is_anomaly'] == True].shape[0]
  # Assert: we have the correct number of anomalies
  assert(num_anomalies == number_of_anomalies)
```

For this scenario, we use the same input dataset. The trickiness behind this dataset is that 2550 and 9000 are clearly outliers but 2550 is so far away from 9000 that if we tried to weight the individual data points and did not cap our statistical tests where we did (3 standard deviations from the mean, 1.5 interquartile ranges from the 25th/75th percentile, and 3 MAD from the median), we might have ended up declaring 2550 as an inlier. As the sensitivity score decreases, we can see the number of corresponding outliers decreases.

The final unit test will ensure that the maximum fraction of anomalies works as expected. We know based on the prior test that when the sensitivity score is 1.0, we get back 17 anomalies. Setting the max fraction of anomalies on a sliding scale, we can see the number of anomalies reported changes. Listing 6-11 contains the code for this test.

Listing 6-11. Controlling sensitivity score and trying different max fractions of anomalies

```
@pytest.mark.parametrize("df_input, max_fraction_anomalies, number_of_
anomalies", [
  (anomalous_sample, 0.0, 0),
  (anomalous_sample, 0.01, 1),
  (anomalous_sample, 0.1, 2),
  (anomalous_sample, 0.2, 3),
  (anomalous_sample, 0.3, 5),
  (anomalous_sample, 0.4, 6),
  (anomalous_sample, 0.5, 8),
  (anomalous_sample, 0.6, 9),
  (anomalous_sample, 0.7, 12),
  (anomalous_sample, 0.8, 12),
  (anomalous_sample, 0.9, 15),
  (anomalous_sample, 1.0, 17)
])
def test_detect_univariate_statistical_sensitivity_affects_anomaly_
count(df_input, max_fraction_anomalies, number_of_anomalies):
  # Arrange
  df = pd.DataFrame(df_input, columns={"value"})
  sensitivity_score = 1.0
  # Act
  (df_out, weights, details) = detect_univariate_statistical(df,
  sensitivity_score, max_fraction_anomalies)
  num_anomalies = df_out[df_out['is_anomaly'] == True].shape[0]
  # Assert: we have the correct number of anomalies
  assert(num_anomalies == number_of_anomalies)
```

These tests give us a better feeling for how the system behaves and will provide additional confidence in the changes we will make in the next chapter.

Updating Integration Tests

The new unit tests we added in the prior section are great, but we also want to make sure that the application as a whole works as expected. That's where new integration tests come into play. For this chapter, we will introduce six new integration tests and perform a minor change to the tests from Chapter 4. First, let's discuss the minor change.

In Chapter 4, we checked for the existence of debugging details by checking the length of the debug_details string. We have subsequently changed the way we write out these details and instead write out a JSON object, so this mechanism needs to change as well. Instead, we now want to check for existence, which means all of the integration test checks will change to the following code: pm.test("Debugging is enabled (debug_details exists): " + body["debug_details"], function () { pm.expect(body["debug_details"].exist); });.

This only affects the tests in which we expect debugging details, so the NoDebug tests will remain the same.

With that change in place, we can add our new requests. The first trio of requests will check sensitivity scores and ensure that different sensitivity scores lead to different numbers of outlier data points. Listing 6-12 gives us one set of tests for this request.

Listing 6-12. The tests that will run with a Postman request

```
pm.test("Status code is 200: " + pm.response.code,
   function () { pm.expect(pm.response.code).to.eql(200); });
pm.test("Response time is acceptable (range 5ms to 7000ms): " +
pm.response.responseTime + "ms",
   function () { pm.expect(_.inRange(pm.response.responseTime, 5, 7000)).
   to.eql(true); });
var body = JSON.parse(responseBody);
pm.test("Number of items returned is correct (eq 17): " +
body["anomalies"].length,
   function () { pm.expect(body["anomalies"].length).to.eql(17); });
 pm.test("Debugging is enabled (debug_details exists): " + body["debug_
   details"],
   function () { pm.expect(body["debug_details"].exist); });
```

```
pm.test("Count of anomalies is correct (eq 2)", () => {
let anomalies = body["anomalies"].filter(a => a.is_anomaly === true);
pm.expect(anomalies.length).to.eql(2);
} );
```

The tests start out the same as before, checking that the response code indicates success, that the response time is acceptable, and that the number of items returned is equal to the number we expect. Then, we put in the new check for debug details and end with a new function. This new function checks to see that the count of anomalies is correct, and it does so by calling the `filter()` function on the `anomalies` array, checking where `is_anomaly` is set to true. This first test uses the default sensitivity score of 50 and max fraction of anomalies of 1.0; therefore, we expect two outliers. A second test decreases the sensitivity score to 10, which reduces the number of outliers to 1. A third test increases the sensitivity score to 75, giving us five outliers.

The second set of requests works off of a new dataset, which contains 125 randomly generated expense report values greater than $0 but less than $40, as well as five expense reports greater than $100. These three tests, by using different combinations of sensitivity score and max fraction of anomalies, lead to differing numbers of outliers.

Figure 6-2 shows the end result of running the entire collection, showing that integration tests for Chapter 4 still succeed, as do the new tests here in Chapter 6.

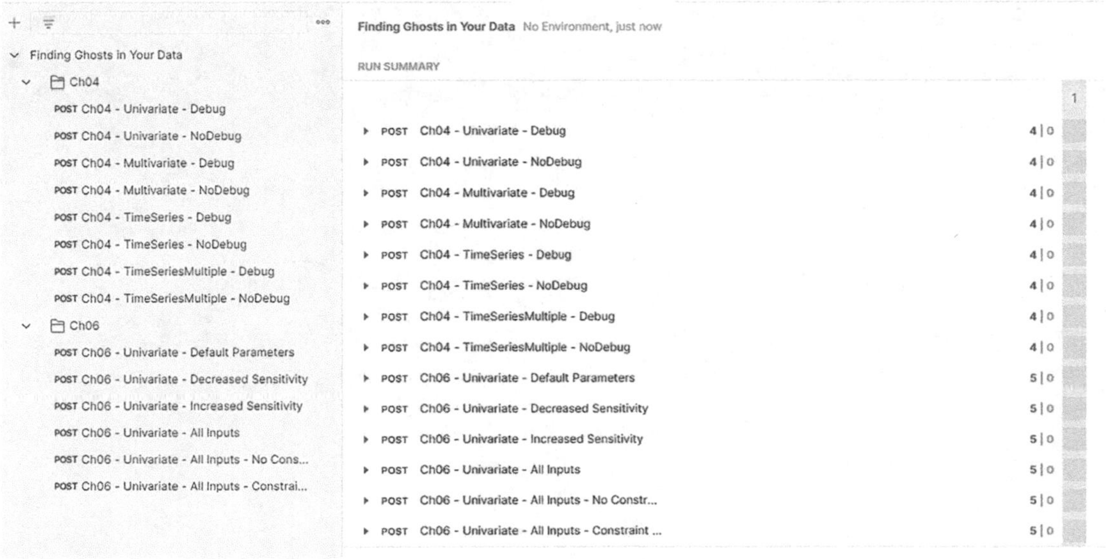

Figure 6-2. *All integration tests passed*

Conclusion

Over the course of this chapter, we turned our outlier detection engine into a functional product. In the next chapter, we will extend the ensemble we created, adding in additional statistical tests, adjusting weights, and aiming to make the univariate detection engine stronger.

## CHAPTER 7

Extending the Ensemble

Chapter 6 provided us with the first three outlier detection tests. In this chapter, we will build upon the prior work and include several additional tests. We will also refactor existing code, rethink a few design choices, and wrap up the core elements of univariate statistical analysis.

Adding New Tests

The three tests we added in the prior chapter are quite useful and might even be enough on their own to solve many common univariate outlier detection problems. If your input data follows a specific shape, however, there are other that can be quite useful in tracking outliers. Specifically, there are several tests we can perform if the data fits closely enough to a normal distribution.

Note   In Chapter 3, we described various distributions, including the normal distribution. As a brief summary, *the* normal distribution has a mean of 0 and a variance of 1. *A* normal distribution is one in which we can, using only addition and multiplication, translate our data to the normal distribution without distorting the relationships between data points. For the rest of this chapter, any reference to normal distributions (even mentions of "the" normal distribution) will be the more generic concept, not the singular normal distribution with a mean of 0 and variance of 1.

© Kevin Feasel 2022
K. Feasel, *Finding Ghosts in Your Data*, https://doi.org/10.1007/978-1-4842-8870-2_7

Checking for Normality

There are several tests that can inform us as to whether a particular dataset appears to follow a normal distribution. The first test we will look at is called the Shapiro-Wilk test. This test, named after Samuel Shapiro and Martin Wilk, tends to be the most accurate at determining whether a given sample of data approximates a normal distribution. In Python, the Shapiro-Wilk test is a function within the `scipy.stats` library.

The second test is called D'Agostino's K-squared test, named after Ralph D'Agostino. This test measures how much *skewness* and *kurtosis* the dataset displays. In statistical terms, the skewness of a sample is a measure of whether the left tail is longer (negative skew), the right tail is longer (positive skew), or both tails are the same length (no skew). Figure 7-1 shows an example of a normal distribution with no skew followed by a log-normal distribution that exhibits positive skew.

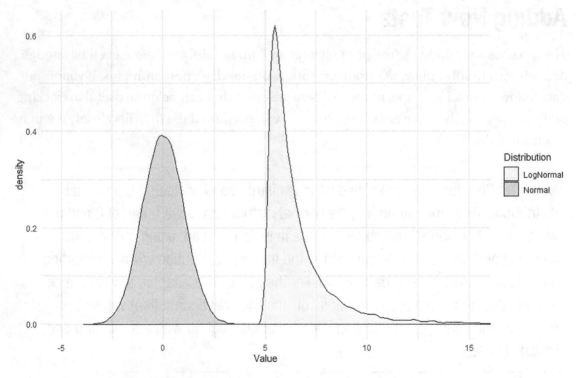

Figure 7-1. *The normal distribution exhibits no skewness. By contrast, the log-normal distribution has a positive skew*

The kurtosis of a dataset indicates the amount of a distribution contained in the tails. The kurtosis of a normal distribution is 3, and as the kurtosis increases, the likelihood of seeing outlier data points increases. Using the datasets from Figure 7-1, the skewness of our normal distribution is 0, and its kurtosis is 2.99. The log-normal dataset has a skewness of 5.34 and a kurtosis of 58.34.

D'Agostino's K-squared test, which is the `normaltest()` function in `scipy.stats`, uses these measures to estimate whether the dataset's distribution is markedly different from the normal distribution.

The third test we will use is the Anderson-Darling test, named after Theodore Anderson and Donald Darling. This test is similar to the Shapiro-Wilk test. It tends not to be quite as accurate as the Shapiro-Wilk test, but in Python, the `anderson()` function performs the test at a variety of significance levels, meaning that you can test at several levels of significance in one go.

Listing 7-1 includes the new import statements for this chapter as well as two of the three function calls: Shapiro-Wilk and D'Agostino's K-squared.

Listing 7-1. Run two basic tests for normality

```
from scipy.stats import shapiro, normaltest, anderson, boxcox
import scikit_posthocs as ph
import math

def check_shapiro(col, alpha=0.05):
  return check_basic_normal_test(col, alpha, "Shaprio-Wilk test", shapiro)

def check_dagostino(col, alpha=0.05):
  return check_basic_normal_test(col, alpha, "D'Agostino's K^2 test",
  normaltest)

def check_basic_normal_test(col, alpha, name, f):
  stat, p = f(col)
  return ( (p > alpha), (f"{name} test, W = {stat}, p = {p}, alpha =
  {alpha}.") )
```

Both of these tests take a similar shape, and so we centralize the code in `check_basic_normal_test()`, which takes four parameters: the values in our DataFrame (`col`), an indicator of just how unlike a normal distribution our result should be before we flag it as non-normal (`alpha`), the name of the function that we want to use for labeling (`name`), and the actual function name (`f`). In Python, we can use a function as an input to another function. When doing so, note that we do not put parentheses around the function name, either in the parameter to `check_basic_normal_test()` or in the calls in `check_shapiro()` or `check_dagostino()`. We save the function call `f(col)` for the body of `check_basic_normal_test()`, in which we call `shapiro()` or `normaltest()` and pass in col as the parameter. These tests calculate a p-value and compare that against our provided `alpha` parameter. If the p-value is greater than alpha, then we assume that the sample could have come from a normal distribution. If the p-value is at or below alpha, we believe that the data is not normal.

Listing 7-2 contains the Anderson-Darling check. Because there is more going on than with the other tests, it does not fit cleanly into the same mold, and therefore, we need to do things a little differently.

Listing 7-2. The Anderson-Darling check for normality

```
def check_anderson(col):
  # Start by assuming normality.
  anderson_normal = True
  return_str = "Anderson-Darling test. "

  result = anderson(col)
  return_str = return_str + f"Result statistic: {result.statistic}"
  for i in range(len(result.critical_values)):
    sl, cv = result.significance_level[i], result.critical_values[i]
    if result.statistic < cv:
      return_str = return_str + f"Significance Level {sl}: Critical Value =
      {cv}, looks normally distributed. "
    else:
      anderson_normal = False
      return_str = return_str + f"Significance Level {sl}: Critical Value =
      {cv}, does NOT look normally distributed! "

  return ( anderson_normal, return_str )
```

Anderson-Darling provides a result statistic and allows you to compare that statistic vs. the critical value for each significance level—that is, each value of alpha. The critical value for each significance level depends on the number of input data points. Figure 7-2 shows what an example looks like for Shapiro-Wilk, D'Agostino, and Anderson-Darling tests against a dataset that passes the normality test.

```
"Initial normality checks": {
    "Shapiro-Wilk": "Shaprio-Wilk test test, W = 0.9701647162437439, p = 0.8923683762550354,
        alpha = 0.05.",
    "D'Agostino": "D'Agostino's K^2 test test, W = 2.02697581498966, p = 0.362950830342156,
        alpha = 0.05.",
    "Anderson-Darling": "Anderson-Darling test.  Result statistic:  0.14110924785979506.
        Significance Level 15.0: Critical Value = 0.501, looks normally distributed.
        Significance Level 10.0: Critical Value = 0.57, looks normally distributed.
        Significance Level 5.0: Critical Value = 0.684, looks normally distributed.
        Significance Level 2.5: Critical Value = 0.798, looks normally distributed.
        Significance Level 1.0: Critical Value = 0.95, looks normally distributed.  "
},
```

Figure 7-2. *The results of three normality checks. Note that the Anderson-Darling significance level scale runs from 0 to 100 rather than 0-1*

In this figure, we can see that the p-values for Shapiro-Wilk and D'Agostino are both well above 0.05, so this dataset passes those normality tests. For Anderson-Darling, the value of the result statistic is 0.141. This number should be below the critical value at each significance level in order for us to consider the dataset as coming from a normal distribution. As we can see, the statistic value of 0.141 is well below the 15% significance level's critical value of 0.501 and therefore below the weaker significance levels as well. By contrast, Figure 7-3 shows an example in which the input dataset does not appear to come from a normal distribution.

```
"Initial normality checks": {
    "Shapiro-Wilk": "Shaprio-Wilk test test, W = 0.35450881719589233, p = 9.431745695565041e-08,
        alpha = 0.05.",
    "D'Agostino": "D'Agostino's K^2 test test, W = 39.95639952541718, p = 2.1065806210913104e-09,
        alpha = 0.05.",
    "Anderson-Darling": "Anderson-Darling test.  Result statistic:  5.039544037522152.
        Significance Level 15.0: Critical Value = 0.501, does NOT look normally distributed!
        Significance Level 10.0: Critical Value = 0.571, does NOT look normally distributed!
        Significance Level 5.0: Critical Value = 0.685, does NOT look normally distributed!
        Significance Level 2.5: Critical Value = 0.799, does NOT look normally distributed!
        Significance Level 1.0: Critical Value = 0.951, does NOT look normally distributed!  "
},
```

Figure 7-3. *A sample that is decidedly not normal*

In this case, the Shapiro-Wilk and D'Agostino p-values are well below 0.05, meaning that both tests indicate that this dataset does not appear to have come from a normal distribution. The Anderson-Darling result statistic has a value greater than 5.0, well above the 0.501 at the 15% significance level and even the 0.951 of the 1% significance level.

Now that we have the three checks in place, we want to build another function that knows when it makes sense to call each method, makes the calls, collates the responses, and makes a determination as to whether the input dataset appears to follow from a normal distribution. Listing 7-3 contains the code for this function.

Listing 7-3. A function that determines whether an input dataset follows from a normal distribution

```
def is_normally_distributed(col):
  alpha = 0.05

  if col.shape[0] < 5000:
    (shapiro_normal, shapiro_exp) = check_shapiro(col, alpha)
  else:
    shapiro_normal = True
    shapiro_exp = f"Shapiro-Wilk test did not run because n >= 5k. n =
    {col.shape[0]}"

  if col.shape[0] >= 8:
    (dagostino_normal, dagostino_exp) = check_dagostino(col, alpha)
  else:
    dagostino_normal = True
    dagostino_exp = f"D'Agostino's test did not run because n < 8. n =
    {col.shape[0]}"

  (anderson_normal, anderson_exp) = check_anderson(col)

  diagnostics = {"Shapiro-Wilk": shapiro_exp, "D'Agostino": dagostino_exp,
  "Anderson-Darling": anderson_exp}
  return (shapiro_normal and dagostino_normal and anderson_normal,
  diagnostics)
```

The function begins by setting an alpha value of 0.05, following our general principle of not having users make choices that require detailed statistical knowledge. Then, if the dataset has fewer than 5000 observations, we run the Shapiro-Wilk test; otherwise, we skip the test. If we have at least eight observations, we can run D'Agostino's K-squared test. Finally, no matter the sample size, we may run the Anderson-Darling test.

When it comes to making a determination based on these three tests, we have to make a decision similar to how we weight our ensembles. We can perform simple voting or weigh some tests greater than others. In this case, we will stick with a simple unanimity rule: if all three tests agree that the dataset is normally distributed, then we will assume that it is; otherwise, we will assume that it is not normally distributed. The diagnostics generate the JSON snippets that we saw in Figures 7-2 and 7-3.

Approaching Normality

Now that we have a check for normality, we can then ask the natural follow-up question: If we need normally distributed data and our dataset is not already normally distributed, is there anything we can do about that? The answer to that question is an emphatic "Probably!"

There are a few techniques for transforming data to look more like it would if it followed from a normal distribution. Perhaps the most popular among these techniques is the Box-Cox transformation, named after George Box and David Cox. Box-Cox transformation takes a variable and transforms it based on the formula in Listing 7-4.

Listing 7-4. The equation behind Box-Cox transformation

$$y = \begin{cases} \log(x) & \text{if } \lambda = 0; \\ \dfrac{(x^{\lambda} - 1)}{\lambda} & \text{otherwise.} \end{cases}$$

Suppose we have some target variable x. This represents the single feature of our univariate dataset. We want to transform the values of x into a variable that follows from a normal distribution, which we'll call y. Given some value of lambda (λ), we can do this by performing one of two operations. If lambda is 0, then we take the log of x, and that becomes y. This works well for log-normally distributed data like we saw in Chapter 3. Otherwise, we have some lambda and raise x to that power, subtract 1 from the value, and divide it by lambda. Lambda can be any value, positive or negative, although common values in Box-Cox transformations range between -5 and 5.

This leads to the next question: How do we know what the correct value of lambda is? Well, the SciPy library has a solution for us: `scipy.stats.boxcox()` takes an optional lambda parameter. If you pass in a value for lambda, the function will use that value to perform Box-Cox transformation. Otherwise, if you do not specify a value for the lambda

123

parameter, the function will determine the optimal value for lambda and return that to you. Even though we get back an optimal value for lambda, there is no guarantee that the transformed data actually follows from a normal distribution, and therefore, we should check the results afterward to see how our transformation fares.

Taking the aforementioned into consideration, Listing 7-5 shows the code we use to perform normalization on an incoming dataset.

Listing 7-5. The function to normalize our input data using the Box-Cox technique

```
def normalize(col):
  l = col.shape[0]
  col_sort = sorted(col)
  col80 = col_sort[ math.floor(.1 * l) + 1 : math.floor(.9 * l) ]
  temp_data, fitted_lambda = boxcox(col80)
  # Now use the fitted lambda on the entire dataset.
  fitted_data = boxcox(col, fitted_lambda)
  return (fitted_data, fitted_lambda)
```

The first thing we do is find the middle 80% of our input data. The rationale behind this is that Box-Cox transformation is quite effective at transforming data. So effective, in fact, that it can smother outliers in the data and make the subsequent tests lose a lot of their value in identifying outlier values. Therefore, we will focus on the central 80% of data, which we typically do not expect to contain many outliers. Running boxcox() on the central 80% of data, we get back a fitted lambda value. We can then pass that value in as a parameter to perform the transformation against our entire dataset, returning the entirety of fitted data as well as the fitted lambda we built from the central 80%.

We make use of this in Listing 7-6, which contains the rules around whether we can normalize the data and how successful that normalization was.

Listing 7-6. The function to perform normalization on a dataset

```
def perform_normalization(base_calculations, df):
  use_fitted_results = False
  fitted_data = None

  (is_naturally_normal, natural_normality_checks) = is_normally_
  distributed(df['value'])
```

```
diagnostics = {"Initial normality checks": natural_normality_checks}
if is_naturally_normal:
  fitted_data = df['value']
  use_fitted_results = True

if ((not is_naturally_normal)
  and base_calculations["min"] < base_calculations["max"]
  and base_calculations["min"] > 0
  and df['value'].shape[0] >= 8):

  (fitted_data, fitted_lambda) = normalize(df['value'])
  (is_fitted_normal, fitted_normality_checks) = is_normally_
  distributed(fitted_data)
  use_fitted_results = True
  diagnostics["Fitted Lambda"] = fitted_lambda
  diagnostics["Fitted normality checks"] = fitted_normality_checks
else:
  has_variance = base_calculations["min"] < base_calculations["max"]
  all_gt_zero = base_calculations["min"] > 0
  enough_observations = df['value'].shape[0] >= 8
  diagnostics["Fitting Status"] = f"Elided for space"

return (use_fitted_results, fitted_data, diagnostics)
```

Listing 7-6 starts by checking to see if the data is already normally distributed. If so, we do not need to perform any sort of transformation and can use the resulting data as is. Otherwise, we perform a Box-Cox transformation if we meet all of the following results: first, the data must not follow from a normal distribution; second, there must be some variance in the data; third, all values in the data must be greater than 0, as we cannot take the logarithm of 0 or a negative number; and fourth, we need at least eight observations before Box-Cox transformation will work.

If we meet all of these criteria, then we call the normalize() function and retrieve the fitted data and lambda. Then, we call is_normally_distributed() a second time on the resulting dataset. Note that we may *still* end up with non-normalized data as a result of this operation due to the fact that we configured the value of lambda based on the central 80% of our dataset and so extreme outliers may still be present in the data after transformation. Therefore, we log the result but do not allow is_normally_ distributed() to prohibit us from moving forward.

In the event that we do not meet all of the relevant criteria, we figure out which criteria are not correct and write that to the diagnostics dictionary. Finally, we return all results to the end user. Now that we have a function to normalize our data should we need it, we can update the run_tests() function to incorporate this, as well as a slew of new tests.

A Framework for New Tests

So far, we have three tests that we want to run regardless of the shape of our data and number of observations. In addition to these three tests, we should add new tests that make sense in particular scenarios. Listing 7-7 shows the outline for the updated run_ tests() function, one which will be capable of running all of our statistical tests.

Listing 7-7. An updated test runner. For the sake of parsimony, most mentions of the diagnostics dictionary have been removed.

```
def run_tests(df):
  base_calculations = perform_statistical_calculations(df['value'])

  (use_fitted_results, fitted_data, normalization_diagnostics) = perform_
  normalization(base_calculations, df)

  # for each test, execute and add a new score
  # Initial tests should NOT use the fitted calculations.
  b = base_calculations
  df['sds'] = [check_sd(val, b["mean"], b["sd"], 3.0) for val in
  df['value']]
  df['mads'] = [check_mad(val, b["median"], b["mad"], 3.0) for val in
  df['value']]
  df['iqrs'] = [check_iqr(val, b["median"], b["p25"], b["p75"], b["iqr"],
  1.5) for val in df['value']]
  tests_run = {
```

```
        "sds": 1,
        "mads": 1,
        "iqrs": 1,
        # Mark these as 0s to start and set them on if we run them.
        "grubbs": 0,
        "gesd": 0,
        "dixon": 0
    }
    # Start off with values of -1. If we run a test, we'll populate it with a
    valid value.
    df['grubbs'] = -1
    df['gesd'] = -1
    df['dixon'] = -1

    if (use_fitted_results):
        df['fitted_value'] = fitted_data
        col = df['fitted_value']
        c = perform_statistical_calculations(col)

        # New tests go here.
    else:
        diagnostics["Extended tests"] = "Did not run extended tests because the
        dataset was not normal and could not be normalized."

    return (df, tests_run, diagnostics)
```

The first thing we do in the function is to perform statistical calculations, retrieving data such as mean, standard deviation, median, and MAD. We have also added the minimum (min), maximum (max), and count of observations (len) to the dictionary, as they will be useful for the upcoming tests. After this, we perform a normalization check and get a response indicating whether we should use the fitted results. Regardless of whether we decide to use the fitted results, we first want to use the original data to perform our checks from Chapter 6. The pragmatic reason is that doing so means we don't need to change as many unit tests in this chapter—transformed data will, by nature of the transformation, not necessarily share the same outlier points as untransformed data.

After running these initial tests, we mark them as having run and add our new tests as not having run. We also create columns in our DataFrame for the new tests. This way, any references to the columns later on are guaranteed to work. Finally, if the use_ fitted_results flag is True, we will set the fitted value to our (potentially) transformed dataset and perform a new set of statistical calculations on the fitted data. Then, we ready ourselves to run any new tests. The following sections cover the three new tests we will introduce, why they are important, and how we will implement them.

Grubbs' Test for Outliers

Grubbs' test for outliers is based on Frank Grubbs' 1969 article, *Procedures for Detecting Outlying Observations in Samples* (Grubbs, 1969). This test outlines a strategy to determine if a given dataset has an outlier. The assumption with this test is that there are no outliers in the dataset, with the alternative being that there is exactly one outlier in the dataset. There is no capacity to detect more than one outlier.

Although this test is quite limited in the number of outliers it produces, it does a good job at finding the value with the largest spread from the mean and checking whether the value is far enough away to matter.

We can run Grubbs' test using the `scikit_posthocs` package in Python. This third-party package includes several outlier detection checks, including Grubbs' test and another test we will use shortly. Calling the test is simple: `ph.outliers_grubbs(col)` is sufficient, where col is the column containing fitted values in our DataFrame. The function call then returns the set of inliers, which means we need to find the difference between our initial dataset and the resulting inlier set to determine if there are any outliers. Listing 7-8 shows one way to do this using set operations in Python.

Listing 7-8. Finding differences between sets to determine if there are outliers

```
def check_grubbs(col):
  out = ph.outliers_grubbs(col)
  return find_differences(col, out)

def find_differences(col, out):
  # Convert column and output to sets to see what's missing.
  # Those are the outliers that we need to report back.
  scol = set(col)
```

```
sout = set(out)
sdiff = scol - sout

res = [0.0 for val in col]
# Find the positions of missing inputs and mark them
# as outliers.
for val in sdiff:
  indexes = col[col == val].index
  for i in indexes: res[i] = 1.0

return res
```

The find_differences() function takes in two datasets: our fitted values (col) and the output from the call to Grubbs' test (out). We convert each of these to a set and then find the set difference—that is, any elements in the fitted values set that do not appear in the results set. Note that duplicate values are not allowed in a set and so all references to a particular outlier value will become outliers. For example, if we have a fitted values dataset of [0, 1, 1, 2, 3, 9000, 9000], the fitted values set becomes { 0, 1, 2, 3, 9000 }. If our output dataset is [0, 1, 1, 2, 3], we convert it to a set as well, making it { 0, 1, 2, 3 }. The set difference between these two is { 9000 }, so every value of 9000 will become an outlier. We do this by creating a new result column that has as many data points as we have elements in the fitted values dataset. Then, for each value in the set difference, we find all references in the fitted values dataset matching that value and mark the matching value in our results dataset to 1.0 to indicate that the data point represents an outlier. We then return the results list to the caller.

The reasons we loop through sdiff, despite there being only one possible element in the set, are twofold. First, this handles cleanly the scenario in which sdiff is an empty set, as then it will not loop at all. Second, we're going to need this function again for the next test, the generalized ESD (or GESD) test.

Generalized ESD Test for Outliers

The generalized extreme Studentized deviate test, otherwise known as GESD or generalized ESD, is the general form of Grubbs' test. Whereas Grubbs' test requires that we have either no outliers or one outlier, generalized ESD only requires that we specify the upper bound of outliers. At this point, you may note that we already keep track of a `max_fraction_anomalies` user input, so we could multiply this fraction by the number of input items and use that to determine the maximum upper bound. This idea makes a lot of sense in principle, but it does not always work out in practice. The reason for this is a concept known as *degrees of freedom.*

Degrees of freedom is a concept in statistics that relates to how many independent values we have in our dataset and therefore how much flexibility we have in finding missing information (in our case, whether a particular data point is an outlier) by varying those values. If we have a dataset containing 100 data points and we estimate that we could have up to five outliers, that leaves us with 95 "known" inliers. Generalized ESD is built off of a distribution called the *t (or student-t) distribution*, whose calculation for degrees of freedom is the number of data points minus one. For this calculation, we take the 95 expected inliers, subtract 1, and end up with 94 degrees of freedom. Ninety-four degrees of freedom to find up to five outliers is certainly sufficient for the job. By contrast, if we allow 100% of data points to be outliers, we have no degrees of freedom and could get an error. Given that our default fraction of outliers is 1.0, using this as the input for generalized ESD is likely going to leave us in trouble.

Instead, let us fall back to a previous rule around outliers and anomalies: something is anomalous in part because it is rare. Ninety percent of our data points cannot be outliers; otherwise, the sheer number of data points would end up making these "outliers" become inliers! Still, we do not know the exact number of outliers in a given dataset, and so we need to make some estimation. The estimation we will make is that we have no more than 1/3 of data points as potential outliers. Along with a minimum requirement of 15 observations, this ensures that we have no fewer than 5 potential outliers for a given dataset and also that we have a minimum of 9 degrees of freedom. As the number of data points in our dataset increases, the degrees of freedom increases correspondingly, and our solution becomes better at spotting outliers.

Listing 7-9 shows how we can find the set of outliers from generalized ESD. It takes advantage of the `find_differences()` function in Listing 7-8. The implementation for generalized ESD comes from the same scikit_posthocs library as Grubbs' test and therefore returns a collection of inlier points. Thus, we need to perform the set difference between input and output to determine which data points this test considers to be outliers.

Listing 7-9. The code to check for outliers using generalized ESD

```
def check_gesd(col, max_num_outliers):
  out = ph.outliers_gesd(col, max_num_outliers)
  return find_differences(col, out)
```

Dixon's Q Test

The final test we will implement in this chapter is Dixon's Q test, named after Wilfrid Dixon. Dixon's Q test is intended to identify outliers in small datasets, with a minimum of three observations and a small maximum. The original equation for Dixon's Q test specified a maximum of ten observations, although Dixon later came up with a variety of tests for different small datasets. We will use the original specification, the one commonly known as the Q test, with a maximum of 25 observations as later research showed that the Q test would hold with datasets of this size. The formula for the Q test is specified in Listing 7-10. We first sort all of the data in x in ascending order, such that x_1 is the smallest value and x_n is the largest.

Listing 7-10. The calculation for Q

$$Q_{exp} = \frac{x_2 - x_1}{x_n - x_1}$$

Once we have calculated the value of Q, we compare it against a table of known critical values for a given confidence level. In our case, we will use the 95% confidence level. Doing so, we check if our expected Q value is greater than or equal to the critical value. If so, the largest data point is an outlier. We can perform a similar test to see if the minimum data point is an outlier. This means that in practice, we can spot up to two outliers with Dixon's Q test: the maximum value and the minimum value.

Listing 7-11 includes the code to perform Dixon's Q test. This test is not available in scikit_posthocs, so we will create it ourselves.

Listing 7-11. The code for Dixon's Q test

```
def check_dixon(col):
  q95 = [0.97, 0.829, 0.71, 0.625, 0.568, 0.526, 0.493, 0.466,
    0.444, 0.426, 0.41, 0.396, 0.384, 0.374, 0.365, 0.356,
    0.349, 0.342, 0.337, 0.331, 0.326, 0.321, 0.317, 0.312,
```

```
   0.308, 0.305, 0.301, 0.29]
 Q95 = {n:q for n, q in zip(range(3, len(q95) + 1), q95)}
 Q_mindiff, Q_maxdiff = (0,0), (0,0)
 sorted_data = sorted(col)

 Q_min = (sorted_data[1] - sorted_data[0])
 try:
   Q_min = Q_min / (sorted_data[-1] - sorted_data[0])
 except ZeroDivisionError:
   pass

 Q_mindiff = (Q_min - Q95[len(col)], sorted_data[0])

 Q_max = abs(sorted_data[-2] - sorted_data[-1])
 try:
   Q_max = Q_max / abs(sorted_data[0] - sorted_data[-1])
 except ZeroDivisionError:
   pass

 Q_maxdiff = (Q_max - Q95[len(col)], sorted_data[-1])

 res = [0.0 for val in col]
 if Q_maxdiff[0] >= 0:
   indexes = col[col == Q_maxdiff[1]].index
   for i in indexes: res[i] = 1.0

 if Q_mindiff[0] >= 0:
   indexes = col[col == Q_mindiff[1]].index
   for i in indexes: res[i] = 1.0

 return res
```

We first create a list of critical values for a given number of observations at a 95% confidence level. We turn this into a dictionary whose key is the number of observations and whose value is the critical value for that many observations. After sorting the data, we calculate the minimum value's Q, following the formula in Listing 7-10. Note that the division is in a try-catch block so that if there is no variance, we do not get an error. Instead, the numerator is guaranteed to be 0. After calculating Q_min, we compare its value to the appropriate critical value and call that Q_mindiff. We can perform a very similar test with the maximum data point, comparing it to the second largest and dividing by the total range of the dataset.

After calculating Q_mindiff and Q_maxdiff, we check to see if either is greater than or equal to zero. A value less than zero means that the data point is an inlier, whereas a value greater than or equal to zero indicates an outlier. Finally, as with the other tests in this chapter, in the event that there are multiple data points with the same value, we mark all such ties as outliers.

Now that we have three tests available to us, we will need to update our calling code to reference these tests.

Calling the Tests

In Listing 7-7, we left a spot for the new tests. Listing 7-12 completes the function call. For the sake of convenience, I have included an if statement and a few preparatory lines from Listing 7-7 to provide more context on where this new code slots in.

Listing 7-12. Add the three new tests.

```
if (use_fitted_results):
  df['fitted_value'] = fitted_data
  col = df['fitted_value']
  c = perform_statistical_calculations(col)
  diagnostics["Fitted calculations"] = c

  if (b['len'] >= 7):
    df['grubbs'] = check_grubbs(col)
    tests_run['grubbs'] = 1
  else:
    diagnostics["Grubbs' Test"] = f"Did not run Grubbs' test because we
    need at least 7 observations but only had {b['len']}."

  if (b['len'] >= 3 and b['len'] <= 25):
    df['dixon'] = check_dixon(col)
    tests_run['dixon'] = 1
  else:
    diagnostics["Dixon's Q Test"] = f"Did not run Dixon's Q test because we
    need between 3 and 25 observations but had {b['len']}."
```

```
  if (b['len'] >= 15):
    max_num_outliers = math.floor(b['len'] / 3)
    df['gesd'] = check_gesd(col, max_num_outliers)
    tests_run['gesd'] = 1
else:
  diagnostics["Extended tests"] = "Did not run extended tests because the
  dataset was not normal and could not be normalized."
```

In order to run Grubbs' test, we need at least 7 observations; running Dixon's Q test requires between 3 and 25 observations; and generalized ESD requires a minimum of 15 observations. If we run a given test, we set the appropriate column in our DataFrame to indicate the results of the test; otherwise, the score will be -1 for a test that did not run.

Because we have new tests, we need updated weights. Keeping the initial tests' weights the same, we will add three new weights: Grubbs' test will have a weight of 0.05, Dixon's Q test a weight of 0.15, and generalized ESD a weight of 0.30. These may seem low considering how much work we put into the chapter, but it is for good reason: Grubbs' test and Dixon's Q test will mark a maximum of one and two outliers, respectively. If there are more anomalies in our data, then they will incorrectly label those data points as inliers. In practice, this provides a good signal for an extreme outlier but a much weaker signal for non-extremes. Generalized ESD has less of a problem with this, as we specify that up to 20% of the data points may be outliers and so its weight is commensurate with the other statistical tests.

Adding these new weights leads to an interesting problem with scoring. In Chapter 6, we made the score a simple summation of weights, allowing that the sum could exceed 1.0. We now have to deal with a new problem: if a test does not run, we don't want to include its weight in the formula. Listing 7-13 shows the new version of the `score_results()` function, which now includes a set of tests we have run.

Listing 7-13. A new way to score results

```
def score_results(df, tests_run, weights):
  tested_weights = {w: weights.get(w, 0) * tests_run.get(w, 0) for w in
  set(weights).union(tests_run)}
  max_weight = sum([tested_weights[w] for w in tested_weights])

  return df.assign(anomaly_score=(
    df['sds'] * tested_weights['sds'] +
```

```
    df['iqrs'] * tested_weights['iqrs'] +
    df['mads'] * tested_weights['mads'] +
    df['grubbs'] * tested_weights['grubbs'] +
    df['gesd'] * tested_weights['gesd'] +
    df['dixon'] * tested_weights['dixon']
  ) / (max_weight * 0.95))
```

The function first creates a dictionary called `tested_weights`, which includes weights for all of the statistical tests we successfully ran, using the value of the weight if we did run the test or 0 otherwise. Then, we sum up the total weight of all tests. Finally, we calculate a new anomaly score as the sum of our individual scored results multiplied by the test weight and divide that by 95% of the maximum weight. The 0.95 multiplier on maximum weight serves a similar purpose to the way we oversubscribed on weights in Chapter 6. The idea here is that even if we have a low sensitivity score, we still want some of the extreme outliers to show up. As a result, our anomaly score for a given data point may go above 1.0, ensuring that the data point always appears as an outlier regardless of the sensitivity level.

With these code changes in place, we are almost complete. Now we need to make a few updates to tests, and we will have our outlier detection process in place.

Updating Tests

A natural result of adding new statistical tests to our ensemble is that some of the expected results in our unit and integration tests will necessarily change. We are adding these new statistical tests precisely because they give us added nuance, so there may be cases in which we had four outliers based on the results of Chapter 6, but now we have five or three outliers with these new tests. As such, we will want to observe the changes that these new statistical tests bring, ensure that the changes are fairly reasonable, and update the unit and integration tests to correspond with our new observations. In addition, we will want to add new tests to cover additional functionality we have added throughout this chapter.

Updating Unit Tests

Some of the tests we created in Chapter 6 will need to change as a result of our work throughout this chapter. The test case with the greatest number of changes is the test that ensures that sensitivity level changes affect the number of anomalies reported. There are several instances in which the number of outliers we saw in Chapter 6 will differ from the number we see in Chapter 7. For example, when we set the sensitivity level to 95 in Chapter 6, we got back 15 outliers. With these new tests in place, we are down to 12 outliers. With the sensitivity level at 85, we went from eight down to five outliers, 75 moves from five to two, and a sensitivity level of 1 goes from one outlier to zero outliers.

This turns out to be a positive outcome, as a human looking at the stream { 1, 1, 2, 2, 3, 3, 4, 4, 5, 5, 6, 7, 8, 9, 10, 2550, 9000 } sees two anomalies: 2550 and 9000. Ideally, our outlier detection engine would only return those two values as outliers. We can see that at least in the range 25–75 (which includes our default of 50), we do in fact get two outliers. There are a few other tests that need minor changes as well for similar reasons.

In addition to these alterations to existing tests, we should create new test cases. Inside the test_univariate.py file, there are several new test datasets. The first is uniform_data, which is a list ranging from 1 to 10. Then, we have normal_data, which is randomly generated data following from a normal distribution with a mean of 50 and a standard deviation of 5. The third dataset, skewed_data, is the normal data list with the exception of four values we multiplied by 1000. Next, we have single_skewed_data, in which only a single value was multiplied by 1000. Finally, larger_uniform_data includes the range from 1 to 50. Listing 7-14 shows a new test, which ensures that normalization calls return the expected results as to whether a given dataset appears to be normal or not. One interesting thing to note is that the Shapiro-Wilk and Anderson-Darling tests tend to evaluate uniformly distributed data as close enough to normal but D'Agostino's K-squared test correctly catches—at least on the larger uniform dataset—that this is a different distribution.

Listing 7-14. A test to ensure that our normalization tests execute as expected on various types of datasets

```
@pytest.mark.parametrize("df_input, function, expected_normal", [
    (uniform_data, check_shapiro, True),
    (normal_data, check_shapiro, True),
    (skewed_data, check_shapiro, False),
    (single_skewed_data, check_shapiro, False),
    (larger_uniform_data, check_shapiro, True),

    (uniform_data, check_dagostino, True),
    (normal_data, check_dagostino, True),
```

```
    (skewed_data, check_dagostino, False),
    (single_skewed_data, check_dagostino, False),
    (larger_uniform_data, check_dagostino, False), # Note that this is
    different from the other tests!

    (uniform_data, check_anderson, True),
    (normal_data, check_anderson, True),
    (skewed_data, check_anderson, False),
    (single_skewed_data, check_anderson, False),
    (larger_uniform_data, check_anderson, True),
])
def test_normalization_call_returns_expected_results(df_input, function,
expected_normal):
    # Arrange
    # Act
    (is_normal, result_str) = function(df_input)
    # Assert: the distribution is/is not normal, based on our expectations.
    assert(expected_normal == is_normal)
```

In this test, we do not need to perform any special preparation work, and so the Arrange section of our test is empty. We can also see that for the most part, there is agreement between these tests, at least for the five extreme cases I chose. We could find other situations in which the tests disagree with one another, but this should prove sufficient for the main purpose of our test.

The natural follow-up to this test is one ensuring that the call to is_normally_ distributed returns the value that we expect. Listing 7-15 provides the code for this test.

Listing 7-15. Combining the three normalization tests provides the expected result

```
@pytest.mark.parametrize("df_input, function, expected_normal", [
    (uniform_data, is_normally_distributed, True),
    (normal_data, is_normally_distributed, True),
    (skewed_data, is_normally_distributed, False),
    (single_skewed_data, is_normally_distributed, False),
    (larger_uniform_data, is_normally_distributed, False),
])
```

```
def test_normalization_call_returns_expected_results(df_input, function,
expected_normal):
  # Arrange
  df = pd.DataFrame(df_input, columns={"value"})
  col = df['value']
  # Act
  (is_normal, result_str) = function(col)
  # Assert: the distribution is/is not normal, based on our expectations.
  assert(expected_normal == is_normal)
```

Because there is disagreement on the larger uniform dataset case, we expect this function to return False. In order for the test to return True, all three statistical tests must be in agreement.

After checking the results of valid tests, we should go further and make sure that the call to perform Box-Cox transformation only occurs when a given dataset meets our minimum criteria: we must have a minimum of eight data points, there must be some variance in the data, and we may not have any values less than or equal to 0 in our dataset. Listing 7-16 includes the code for this unit test, as well as a few new datasets for our test cases.

Listing 7-16. Ensure that we only perform normalization on valid datasets

```
@pytest.mark.parametrize("df_input, should_normalize", [
  (uniform_data, True), # Is naturally normal (enough)
  (normal_data, True), # Is naturally normal
  (skewed_data, True),
  (single_skewed_data, True),
  # Should normalize because D'Agostino is false.
  (larger_uniform_data, True),
  # Not enough datapoints to normalize.
  ([1,2,3], False),
  # No variance in the data.
  ([1,1,1,1,1,1,1,1,1,1,1,1,1,1,1], False),
  # Has a negative value--all values must be > 0.
  ([100,20,3,40,500,6000,70,800,9,10,11,12,13,-1], False),
  # Has a zero value--all values must be > 0.
```

```
    ([100,20,3,40,500,6000,70,800,0,10,11,12,13,14], False),
])
def test_perform_normalization_only_works_on_valid_datasets(df_input,
should_normalize):
    # Arrange
    df = pd.DataFrame(df_input, columns={"value"})
    base_calculations = perform_statistical_calculations(df['value'])
    # Act
    (did_normalize, fitted_data, normalization_diagnostics) = perform_
    normalization(base_calculations, df)
    # Assert: the distribution is/is not normal, based on our expectations.
    assert(should_normalize == did_normalize)
```

The comments next to each test case explain why we expect the results we get. Listing 7-17 shows the final test we add, which indicates whether a given statistical test may run based on the dataset we pass in. For example, Grubbs' test requires a minimum of seven data points, so if we have fewer than seven, we cannot run that check. The specific test cases are in the test_univariate.py file, with Listing 7-17 showing only the calling code.

Listing 7-17. The code to determine if we run a specific statistical test

```
def test_normalization_required_for_certain_tests(df_input, test_name,
test_should_run):
    # Arrange
    df = pd.DataFrame(df_input, columns={"value"})
    # Act
    (df_tested, tests_run, diagnostics) = run_tests(df)
    # Assert: the distribution is/is not normal, based on our expectations.
    assert(test_should_run == diagnostics["Tests Run"][test_name])
```

We can confirm whether a test ran by reading through the diagnostics dictionary, looking inside the Tests Run object for a particular test name.

After creating all of the relevant tests, we now have 80 test cases covering a variety of scenarios. We will add more test cases and tests as we expand the product, but this will do for now. Instead, we shall update some existing integration tests and then add new ones for Chapter 7.

Updating Integration Tests

Just like with the unit tests, we will need to perform some minor alterations to integration tests. For example, the integration test named Ch06 – Univariate – All Inputs – No Constraint returned 14 outliers in Chapter 6, but with the new tests, that number drops to 8.

Aside from these changes, we will add three more integration tests indicating whether normalization succeeds, fails, or is unnecessary. Listing 7-18 shows the test cases for a Postman-based integration test in which we expect normalization to succeed.

Listing 7-18. We expect normalization to take place for this call.

```
pm.test("Status code is 200: " + pm.response.code, function () {
pm.expect(pm.response.code).to.eql(200); });
pm.test("Response time is acceptable (range 5ms to 7000ms): " +
pm.response.responseTime + "ms", function () { pm.expect(_.inRange(pm.
response.responseTime, 5, 7000)).to.eql(true); });
var body = JSON.parse(responseBody);
pm.test("Number of items returned is correct (eq 17): " +
body["anomalies"].length, function () { pm.expect(body["anomalies"].
length).to.eql(17); });
pm.test("Debugging is enabled (debug_details exists): " + body["debug_
details"], function () { pm.expect(body["debug_details"].exist); });
pm.test("Count of anomalies is correct (eq 2)", () => {
  let anomalies = body["anomalies"].filter(a => a.is_anomaly === true);
  pm.expect(anomalies.length).to.eql(2);
} );
pm.test("We have fitted results (Fitted Lambda exists): " + body["debug_
details"]["Test diagnostics"]["Fitted Lambda"], function () {
pm.expect(body["debug_details"]["Test diagnostics"]["Fitted Lambda"].
exist); });
```

The final test, in which we check for the existence of a fitted lambda (indicating that we performed Box-Cox transformation), is the way in which we indicate that normalization did occur as we expected. We have similar checks in the other tests as well.

Multi-peaked Data

Now that our univariate anomaly detection engine is in place, we are nearly ready to wrap up this chapter. Before we do so, however, I want to cover one more topic: multi-peaked data.

A Hidden Assumption

Thus far, we have worked from a fairly straightforward and reasonable assumption: our data is approximately normally distributed. Furthermore, we assume that our data is single peaked. Figure 7-4 is a reprint of Figure 7-1, which we saw at the beginning of the chapter.

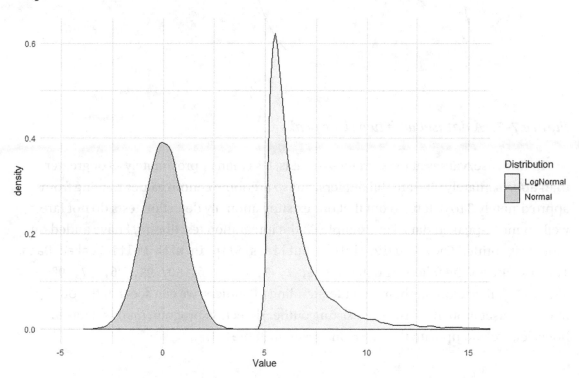

Figure 7-4. *Two distributions of data, each of which has a single peak*

This figure demonstrates two single-peaked curves. There is a clear central value in which density is highest. By contrast, Figure 7-5 shows an example of a multi-peaked curve with several competing groups of values.

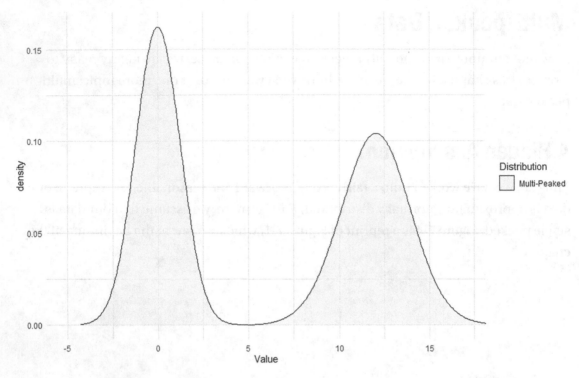

Figure 7-5. *A dataset with two clear peaks*

In this case, values at the extremes—that is, less than approximately -3 or greater than approximately 20—are still outliers, but so are some central values ranging from approximately 3 to 8. It turns out that our existing anomaly detection tests do not fare well on multi-peaked data. For example, in the integration test library, I have added one more test entitled Ch07 - Univariate - Outliers Slip in with Multi-Peaked Data. This test checks the following dataset: { 1, 2, 3, 4, 5, 6, 50, 95, 96, 97, 98, 99, 100 }. Based on our human understanding of outliers, we can see that the outlier in this dataset is 50. If we run our existing outlier detection program on this dataset, however, we end up with the surprising result in Table 7-1.

Table 7-1. *The results of a test with multi-peaked data*

Value	Anomaly Score
1	0.169
2	0.164
3	0.158
4	0.153
5	0.149
6	0.146
50	0.001
95	0.148
96	0.151
97	0.155
98	0.160
99	0.165
100	0.171

Not only does 50 not get marked as an outlier, it's supposedly the *least* likely data point to be an outlier! The reason for this is that our statistical tests are focused around the mean and median, both of which are (approximately) 50 in this case.

The Solution: A Sneak Peek

Now that we see what the problem is, we can come up with a way to solve this issue. The kernel of understanding we need for a solution is that we should not necessarily think of the data as a single-peaked set of values following from an approximately normal distribution. Instead, we may end up with cases in which we have groups of data, which we will call *clusters*. Here, we have two clusters of values: one from 1 to 6 and the other from 95 to 100. We will have much more to say about clustering in Part III of the book.

Conclusion

In this chapter, we extended our capability to perform univariate anomaly detection. We learned how to check if an input dataset appears to follow from a normal distribution and how to perform Box-Cox transformation on the data if it does not follow a normal distribution, and incorporated additional tests that expect normalized data in order to work. Our univariate anomaly detector has grown to nearly 400 lines of code and has become better at picking out outliers, returning fewer spurious outliers as we increase the sensitivity level.

In the next chapter, we will take a step back from this code base. We will build a small companion project to visualize our results, giving us an easier method to see outputs than reading through hundreds of lines of JSON output.

CHAPTER 8

Visualize the Results

Over the course of the last two chapters, we have interacted with our outlier detection service: we have built and extended functionality, created and updated tests, and diagnosed issues by reading a lot of JSON output. In this chapter, we will provide a more user-friendly method for integrating with the outlier detector. This will come in the form of a web application.

Building a Plan

The first thing to keep in mind as we work through this solution is that the web service is a *consumer* of our outlier detection service, not a part of the service itself. As such, we will want to build an independent application. This application will make calls to our detection service and display the results in a way that will make it easier for users to see the results. Because we want to minimize the amount of application development work here, we will emphasize web applications rather than desktop apps. Also, this visualization tool is not something we intend to make available to the general public—at least not the version we're going to use in this book. The principles could drive a more complete product, but for our purposes, we simply want something easier to read than a block of JSON.

With those basic principles in mind, let's get into the information we currently have available and how best we can display that information.

What Do We Want to Show?

There are several things we will want to show on a simple application. First, we want to provide a way for users to enter data. We should also provide enough guidance to ensure that users know how we need that data. That way, if they develop production-quality applications that hit our outlier detection service, they will know the structure of what

145

© Kevin Feasel 2022
K. Feasel, *Finding Ghosts in Your Data*, https://doi.org/10.1007/978-1-4842-8870-2_8

they need to send in. We should also provide some capability to choose the method of detection. So far, we have focused entirely on univariate outlier detection, but that will soon change and we should be prepared for that change. We also should allow users to choose whether they wish to see debug data.

As far as outputs go, we want to do something better than simply regurgitate the JSON we receive. Instead, we can provide a graphical interpretation of our data. With univariate outlier detection, the solution is fairly easy: our single input variable, `value`, becomes the X axis of a scatter chart, and the anomaly score represents the Y axis. We could also write out the data points in a tabular format. The combination of these two facilities will make it easier for a human to interpret the data and adjudge whether our detected outliers are actually anomalies or if they are merely noise. It would also be nice to display debugging details in a way that makes logical sense. We might show the different collections of data, such as debugging weights, tests run, and the different calculations separately. That way, a person can quickly scan to the desired section and see results. If a debugging section is not relevant, such as the fitted calculations when we do not perform normalization, we can either display nothing at all or show a message explaining why a given section is not relevant.

How Do We Want to Show It?

We have a lot of choices in how we can build a web application to use our outlier detection service. Because we are going to call a REST API, we do not need to write the client application in Python and can instead choose any language that supports calls to REST APIs. For the sake of simplicity, we will in fact develop this application in Python, but if you are more familiar with Java, F#, C#, R, or some other language, building your own application to call this service can be a useful exercise.

Sticking to Python, we have quite a few options for application development. One option is to build a proper website using a framework such as Django, TurboGears, or CherryPy. This would give us the ability to design a detailed set of interactions, and it might be the best route for a complete web application that happens to integrate outlier detection. For this chapter, however, such a site would be overkill for us. Fortunately, there are other options available that are more suited for the immediate task.

One such option is called Dash and is based on a visualization library called Plotly. Dash provides us a great dashboard development experience, making it a natural choice for production application development. There is, however, a bit of a learning curve with Dash, so we will instead use a library called Streamlit. Streamlit exists to provide data scientists an easy way to spin up a simple website to work with data and models. Unlike Dash, Streamlit is not a "production-quality" service, so we would not want to expose this to the general public. For our purposes in building an internal tool, however, Streamlit will be an excellent choice.

Developing a Visualization App

To install Streamlit, run `pip install streamlit` at the command line. This will install all of the necessary libraries and dependencies for us. Next, create a folder called `\web\` in the `\src\` folder, and inside there, create `site.py`. We will use this separate folder for managing the Streamlit website.

Getting Started with Streamlit

Streamlit has a series of built-in functions intended to make it easy to display text and data on screen. The `write()` function allows us to display contents that we create using Markdown formatting. In addition to that, there are a variety of input fields like check boxes, radio buttons, text boxes, and text areas, allowing us to receive user input. The biggest quality-of-life feature in Streamlit is its ease of displaying Pandas DataFrames and graphs from a variety of libraries like Plotly, Seaborn, and matplotlib. Streamlit has a great cheat sheet at `https://docs.streamlit.io/library/cheatsheet`, as well as a good deal of API documentation on their website.

Because we can keep this site simple, all of our website code will be in the single Python file. Therefore, we will need a main function. Listing 8-1 shows the basic shell of our website.

Listing 8-1. The shell of a Streamlit app

```
import streamlit as st
import requests
import pandas as pd
import json
import plotly.express as px

st.set_page_config(layout="wide")
```

```
def main():
  st.write("TODO: fill in.")

if __name__ == "__main__":
  main()
```

We first import several libraries, including plotly.express to write out visuals. Then we set the page layout to a wide setting. By default, Streamlit uses a small percentage of total screen space, especially on large monitors. Setting the layout to "wide" means that we will take up the entire width of the browser, which is much more convenient for our purposes.

After that, we create a main function and populate it with a stub message indicating that we need to fill in the details. Finally, we execute the main function to display our page.

Once you have everything set up, open a console window to the \src\web\ folder and run `streamlit run site.py`. This will open up a new browser window and show our simple website. With this structure in place, let's expand on it a bit.

Building the Initial Screen

The first thing we want to do is to create an input screen for users, allowing them to select an outlier detection method (though so far, we only have the univariate method implemented), specify a valid sensitivity score, specify a max fraction of anomalies, tell us whether we should run in debug mode or not, and send us a valid input dataset in JSON format. Because we cannot expect users to intuit what we expect in terms of inputs, we should provide some instructions in the form of a sample dataset. Listing 8-2 shows the code needed to set this up.

Listing 8-2. The code to build a user input section

```
server_url = "http://localhost/detect"
method = st.selectbox(label="Choose the method you wish to use.", options =
("univariate", "multivariate", "timeseries/single", "timeseries/multiple"))
sensitivity_score = st.slider(label = "Choose a sensitivity score.", min_
value=1, max_value=100, value=50)
max_fraction_anomalies = st.slider(label = "Choose a max fraction of
anomalies.", min_value=0.01, max_value=1.0, value=0.1)
```

```
debug = st.checkbox(label="Run in Debug mode?")

if method == "univariate":
  starting_data_set = """[
  {"key": "1","value": 1},
  {"key": "2", "value": 2},
  {"key": "3", "value": 3},
  {"key": "4", "value": 4},
  {"key": "5", "value": 5},
  {"key": "6", "value": 6},
  {"key": "8", "value": 95}
]
  """
# Omitting elif blocks for simplicity.
else:
  starting_data_set = "Select a method."
input_data = st.text_area(label = "Data to process (in JSON format):",
value=starting_data_set, height=500)

if st.button(label="Detect!"):
  st.write("TODO: finalize details once we input data.")
```

First, we set the server URL. This is the location of our outlier detection program, which we will have running locally on port 80.

Note   In order for this Streamlit website to work, you will need to have the uvicorn and the FastAPI service running. To do so, the command `uvicorn app.main:app --host 0.0.0.0 --port 80` will host the API on port 80 of your local machine, assuming you do not have another service hosted on that port.

Next, we want to allow users to choose the type of outlier detection they wish to perform. Because we have only implemented univariate outlier detection to this point, we will set that as the default. Once we have implemented other techniques, we can come back to this site and make updates as needed.

The second user input pertains to the sensitivity score. This score should range from 1 to 100, so we will add a slider to allow users to choose. The third user input allows the caller to set the max fraction of anomalies to a value ranging from 0.01 to 1.0, with a default of 0.1. After that, we have a simple checkbox to determine if we should run the command in debug mode or not. Then, we generate a text area using the `text_area()` function. This text area will allow users to type or paste in blocks of JSON matching the appropriate input format. To make it easier for users to know what that input format should look like, we provide as a default value the contents of the `starting_data_set` variable. The idea here is that we create a very simple but properly formatted input dataset for users, allowing brand new users a chance to run the command and get immediate feedback.

Finally, we place a button on the page and label it with "Detect!" When the user clicks the button, we will perform the call to our service and render data on-screen. Figure 8-1 shows the current status of the application.

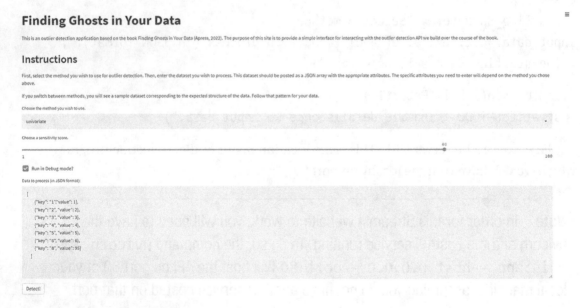

Figure 8-1. *After the first phase of application development, we have user controls in place*

We can see how easy it is to build a simple (if rudimentary) app with Streamlit: with just a few lines of Python code, we have a form that accepts user inputs and even performs an action once we click a button. In the next section, we will make that button do much more.

Displaying Results and Details

At the end of the prior section, our button wrote a "TODO" message to screen. We will now replace that code with a call to our outlier detection service. We could make this call directly in the main() function, but there are two benefits to breaking this call out into its own function. First, our main function will become a little unwieldy because we are building a user interface with it. Therefore, breaking out operational calls will save us some troubleshooting headaches later. Second, as we see in Listing 8-3, having a separate function allows us to tag that function with a Streamlit annotation called cache.

Listing 8-3. The function to call the outlier detection service, including a cache annotation

```
@st.cache
def process(server_url, method, sensitivity_score, max_fraction_anomalies,
debug, input_data_set):
  full_server_url = f"{server_url}/{method}?sensitivity_score={sensitivity_
  score}&max_fraction_anomalies={max_fraction_anomalies}&debug={debug}"
  r = requests.post(
    full_server_url,
    data=input_data_set,
    headers={"Content-Type": "application/json"}
  )
  return r
```

The cache annotation *memoizes* function executions. This is a fancy way of saying that for each unique set of inputs, Streamlit will save the resulting output of the function call. Then, if we make that call again while the application is running, it will retrieve the results from its cache rather than calling our outlier detection service. This works fine for us because we expect the service to return the same results each time we call it for a given method, sensitivity score, max fraction of anomalies, debug status, and input dataset.

Having our process() function in place, we may return to the main() function and replace the "TODO" text with the code in Listing 8-4. This code displays the results in an easy-to-understand format.

151

Listing 8-4. Display the results of univariate anomaly detection.

```
resp = process(server_url, method, sensitivity_score, max_fraction_
anomalies, debug, input_data)
res = json.loads(resp.content)
df = pd.DataFrame(res['anomalies'])

st.header('Anomaly score per data point')
colors = {True: 'red', False: 'blue'}
g = px.scatter(df, x=df["value"], y=df["anomaly_score"], color=df["is_
anomaly"], color_discrete_map=colors,
  symbol=df["is_anomaly"], symbol_sequence=['square', 'circle'],
  hover_data=["sds", "mads", "iqrs", "grubbs", "gesd", "dixon"])
st.plotly_chart(g, use_container_width=True)

tbl = df[['key', 'value', 'anomaly_score', 'is_anomaly', 'sds', 'mads',
'iqrs', 'grubbs', 'gesd', 'dixon']]
st.write(tbl)
```

The first thing we do is call the outlier detection service, giving us back a response. Then, we take the contents of our response and convert it to a JSON object. Inside this JSON object, we have a list called anomalies, which we can convert into a Pandas DataFrame.

We want to display this data in two ways: in graphical format as well as tabular. The graphical format will show the data as a scatter plot, using the value as the X axis and anomaly score as the Y axis. We will mark anomalies as red squares and non-anomalies as blue circles, making it easy to differentiate between the two, even for users with color vision deficiency.

Caution   When choosing colors, always be mindful of color vision deficiency (CVD, also colloquially referred to as color blindness), which affects approximately 8% of men and 0.5% of women. Color vision deficiency may make it difficult to differentiate certain color combinations, such as red and green. Use a tool like Coblis (`www.color-blindness.com/coblis-color-blindness-simulator/`) to see how your images will appear to those with various types of color vision deficiency. In this case, the colors for red and blue do not overlap for any of the common forms of CVD, but they are hard to differentiate in monochromatic media, such as in the pages of this book! Therefore, we also use shape as an identifier, making it easier to see differences.

By default, hovering over a data point will show the values for the X and Y axes—that is, value and anomaly score, respectively—as well as the indicator for color, which is whether the data point is an outlier. We might want to include additional information, such as the results of each test. To do that, we include the columns we wish to include in the hover_data list. We might also think about setting the X axis to a logarithmic scale by adding log_x=True as a parameter. This would allow us to see very large differences in data without squishing smaller values together too much, but it does come with one significant downside: all of the values on the X axis must be positive; if you have negative values, they will not appear on the chart if you set the X axis to logarithmic scale. For this reason, we will not display X axis values on a logarithmic scale, though if you know your input datasets will always include numbers greater than zero, you might wish to think about making this change.

After the chart, we also include a table with our data points. That way, a human can scan through the results and compare items side by side without needing to hover over each element. Figure 8-2 shows our progress so far.

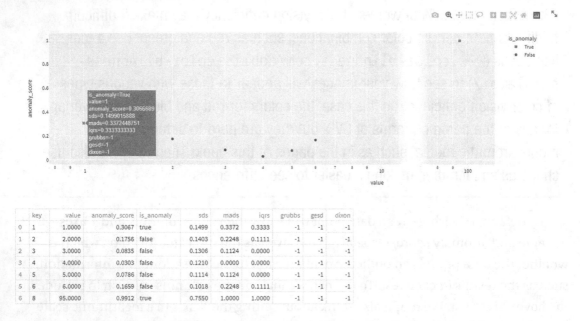

Detect!

Anomaly score per data point

	key	value	anomaly_score	is_anomaly	sds	mads	iqrs	grubbs	gesd	dixon
0	1	1.0000	0.3067	true	0.1499	0.3372	0.3333	-1	-1	-1
1	2	2.0000	0.1756	false	0.1403	0.2248	0.1111	-1	-1	-1
2	3	3.0000	0.0835	false	0.1306	0.1124	0.0000	-1	-1	-1
3	4	4.0000	0.0303	false	0.1210	0.0000	0.0000	-1	-1	-1
4	5	5.0000	0.0786	false	0.1114	0.1124	0.0000	-1	-1	-1
5	6	6.0000	0.1659	false	0.1018	0.2248	0.1111	-1	-1	-1
6	8	95.0000	0.9912	true	0.7550	1.0000	1.0000	-1	-1	-1

Figure 8-2. The anomaly score per data point. Note that a section of the graph without any data points was removed in order to make the relevant sections of the screen easier to read. The removed section was replaced with a small gap to make this removal clear. In this example, we also use a log scale for the X axis to make it easier to see results in print

The final thing we will need to add is debugging details. Unlike the set of outliers, our debugging details will not cleanly fit into a Pandas DataFrame, as it is a complex, nested document. We can, however, break the details up into several sections. In order to maximize our available visible space, we can write the results out to multiple columns using the st.columns() function. Listing 8-5 shows the first two blocks of data: debug weights and the tests we have run.

Listing 8-5. The first two blocks of information we write out when debugging

```
if debug:
  col11, col12 = st.columns(2)

  with col11:
    st.header('Debug weights')
```

```
    st.write(res['debug_weights'])

with col12:
  st.header("Tests Run")
  st.write(res['debug_details']['Test diagnostics']['Tests Run'])
  if "Extended tests" in res['debug_details']['Test diagnostics']:
    st.write(res['debug_details']['Test diagnostics']['Extended tests'])

# ...
```

We know that some of the blocks, such as debug weights, will always be available. Other blocks, such as whether we ran extended tests or not, will depend upon the specific dataset we sent in. Therefore, we will need conditional statements to ensure that the information exists before attempting to write it out; otherwise, Streamlit will emit an error message on the page.

After handling debug weights and test runs, we also write out base calculations, fitted calculations, initial normality checks, fitted normality checks, and finally full debug details in its entirety. Figure 8-3 shows the results of this code.

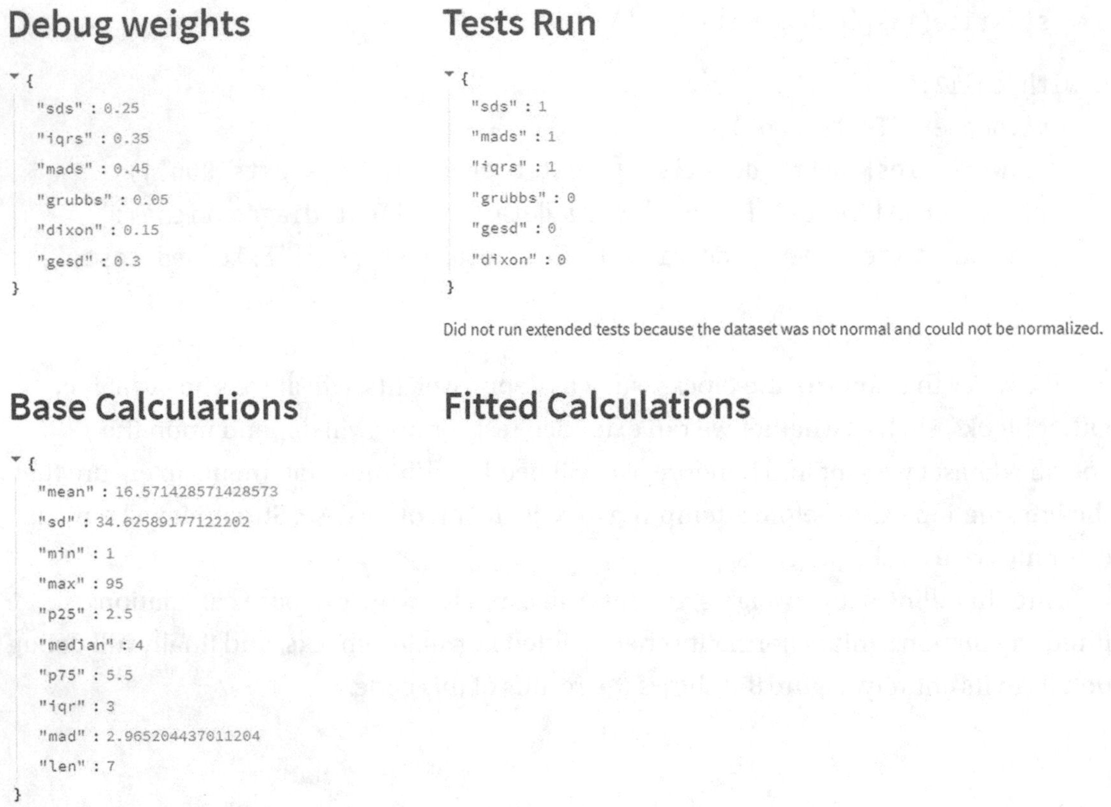

Debug weights

```
▼ {
    "sds" : 0.25
    "iqrs" : 0.35
    "mads" : 0.45
    "grubbs" : 0.05
    "dixon" : 0.15
    "gesd" : 0.3
  }
```

Tests Run

```
▼ {
    "sds" : 1
    "mads" : 1
    "iqrs" : 1
    "grubbs" : 0
    "gesd" : 0
    "dixon" : 0
  }
```

Did not run extended tests because the dataset was not normal and could not be normalized.

Base Calculations

```
▼ {
    "mean" : 16.571428571428573
    "sd" : 34.62589177122202
    "min" : 1
    "max" : 95
    "p25" : 2.5
    "median" : 4
    "p75" : 5.5
    "iqr" : 3
    "mad" : 2.965204437011204
    "len" : 7
  }
```

Fitted Calculations

Figure 8-3. *Diagnostic information, which we have broken into its core components*

Streamlit has the ability to understand Python dictionaries and write them out in a colorful, interactive format. The drop-down markers on the top left of each dictionary allow us to hide and display the dictionary contents, something which is particularly useful in complex documents like the full debug details. One thing we cannot do is easily format the results. This means that the initial and fitted normality checks, specifically the Anderson-Darling test, will be a little more difficult to read than is ideal. Nonetheless, in fewer than 150 lines of Python code, we have a functional website that we can use to continue testing our outlier detection service.

Conclusion

Over the course of this chapter, we have created a simple website that we can use to display the results of our outlier detection service. We will need to expand and revise this service over the upcoming chapters as we implement new outlier detection techniques. Over the course of the next part of this book, we will implement one of these techniques: multivariate outlier detection. Further, we will use some of the insights from that process to improve univariate outlier detection as well.

PART III

Multivariate Anomaly Detection

Univariate outlier detection was the theme for Part II of the book; in this part, we will switch gears to multivariate outlier detection. Chapter 9 serves as the jumping-off point, as we cover the idea of clusters and see how clustering techniques can help us find outliers. We also wrap up univariate outlier detection by incorporating a univariate clustering technique into our existing ensemble. In Chapter 10, we introduce one of the multivariate clustering techniques we will use: Connectivity-Based Outlier Factor (COF). Chapter 11 brings us the other multivariate clustering technique: Local Correlation Integral (LOCI). Chapter 12 provides an important reminder that clustering techniques are not the only useful tools for multivariate analysis, as we review and implement Copula-Based Outlier Detection (COPOD).

## CHAPTER 9

Clustering and Anomalies

This first chapter of Part III also serves as a bridge between Part II, univariate analysis, and Part III, multivariate analysis. In this chapter, we will look at how we can use clustering techniques to solve the problem we ended Chapter 7 with: we can intuitively understand that in a dataset with values { 1, 2, 3, 50, 97, 98, 99 }, the value 50 is an outlier but our outlier detection engine steadfastly refuses to believe it.

This chapter will begin with a description of the concept of clustering. Then, we will look at two very popular techniques for clustering: k-means clustering and k-nearest neighbors. After this, we will cover areas in which clustering techniques make sense, as well as the problems they can introduce. Once we have a solid foundation around clustering techniques, we will add one more method to the univariate anomaly detection process: Gaussian mixture. We will learn how this algorithm works and how we can incorporate its insights into our existing anomaly detector without introducing too much chaos. Finally, we will look at three alternative approaches for multivariate anomaly detection that we will not use in our outlier detection process but still bear mentioning.

What Is Clustering?

The key insight behind clustering is that we typically see data points group together, especially as we introduce more and more variables. As an example of this, take the venerable iris dataset, which Edgar Anderson originally collected in 1935. Figure 9-1 shows three classes of iris and the values for petal length and petal width.

© Kevin Feasel 2022
K. Feasel, *Finding Ghosts in Your Data*, https://doi.org/10.1007/978-1-4842-8870-2_9

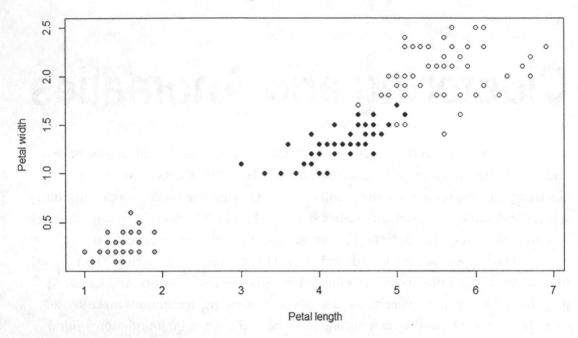

Figure 9-1. *Two variables from the iris dataset show clusters of values*

We can see that there are at least two clear clusters of data: one with petal lengths of 1–2 cm and widths of 0–0.5 cm and a broader cluster containing petal lengths of 3–7 cm and widths of 1–2.5 cm. We do not see any petals with widths over 1 cm but lengths less than 3 cm, nor do we see long and thin petals.

I mention "at least" two clusters because we can make a reasonable argument that there are three clusters here, one for each class of iris in the dataset. Although two classes do have a fair amount of overlap, our eyes can differentiate the two and come up with a fuzzy boundary between the two.

Now that we have an intuitive understanding of the idea of clustering, let's dive into a few terms before looking at different approaches to clustering.

Common Cluster Terminology

We wish to perform *cluster analysis*, which is grouping together observations in some way that maximizes *intracluster similarity* and minimizes *intercluster similarity*. Another way of stating this is that we want the size of each cluster to be as small as possible and for the clusters to be as far apart from each other as possible.

Typically, we draw out a cluster starting from a *centroid* or center point of the cluster. This centroid need not be a specific data point but may instead lie between data points.

One measure we use to calculate intracluster similarity is called *within-cluster sum of squares* (WCSS). We calculate WCSS for each cluster by summing the square of the Euclidean distance between a given point and the centroid. Listing 9-1 shows this formally, assuming a cluster called C.

Listing 9-1. The formula to calculate the within-cluster sum of squares

$$WCSS = \sum_{x \in C} \|x - \mu\|^2$$

The measure for intercluster similarity is called *between-cluster sum of squares* (BCSS). We can calculate BCSS by calculating the total sum of squares in the dataset and subtracting the within-cluster sum of squares; this will leave the between-cluster sum of squares.

Now that we have the basics of necessary terminology behind us, we can begin looking at clustering algorithms.

K-Means Clustering

K-means clustering is a common approach to clustering because it is reasonably fast and very easy to understand. We will start with Figure 9-2, which represents a set of points across two dimensions, which we map as the X and Y axes of a scatter chart.

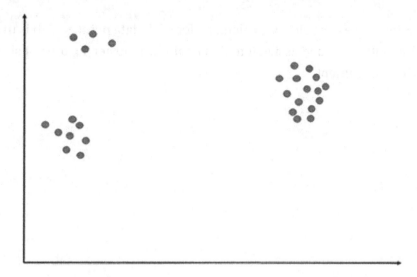

Figure 9-2. *An example of data that exhibits clustering behavior*

We can clearly see from this data that there are three clusters. The way k-means clustering works, we need to determine the number of clusters we believe will exist in the data. This makes our gestalt-driven observations potentially quite useful. If you do not know the number of clusters or have not seen the data before analysis and therefore cannot generate any expectations, you could try this algorithm with different numbers of clusters, as we will see. In this case, after observing the data visually, I may determine that there are three clusters. During the first phase, the algorithm arbitrarily assigns locations for the cluster centroids, which we see in Figure 9-3.

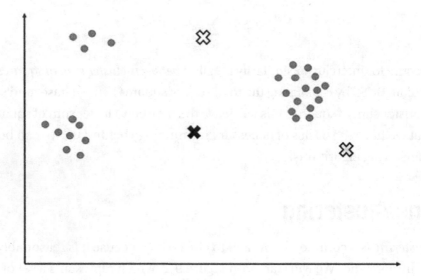

Figure 9-3. *Three new cluster centroids spring into existence*

After creating these centroids, we calculate, for each data point, which is the closest centroid to that data point and assign it to that centroid's cluster. Figure 9-4 shows the outcome of that assignment.

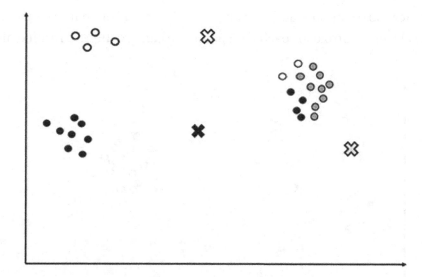

Figure 9-4. *The results of the initial cluster assignment*

After this initial assignment, we can see that the results do not at all match our intuition. That is okay, however, because the next step in the process is to recalculate the centroid locations. The new centroid location will be the center of each assigned cluster, meaning that the black centroid will be the average value of all of the black data points, the white centroid will be the average value of all of the white data points, and the gray centroid will be the average value of all of the gray data points. Figure 9-5 shows these new positions.

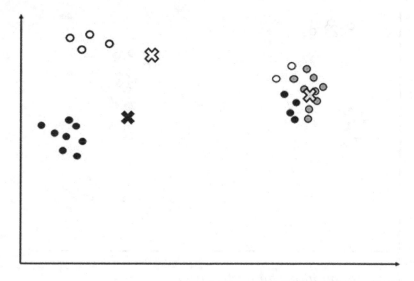

Figure 9-5. *The new locations for our cluster centroids*

With the new positions in place, we can perform our cluster analysis once more, choosing the closest centroid for each data point. We can see the end result of this in Figure 9-6.

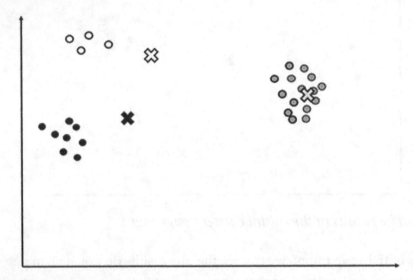

Figure 9-6. *The next round of cluster assignment makes things much clearer*

We repeat this process of calculating the centroid location and assigning values until we reach a state like Figure 9-7, in which the centroid positions do not move (beyond a predefined tolerance level) and we do not see any change in cluster assignments.

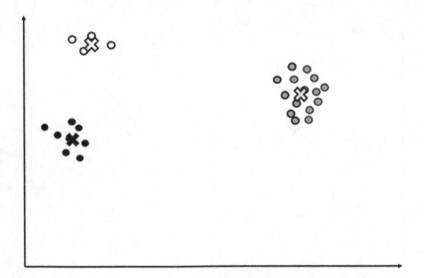

Figure 9-7. *The final positions for each cluster*

At this point, we have reached *convergence*: the clusters are now in a stable state in which further iterations will affect neither the composition of each cluster nor the position of each centroid. This example was intentionally obvious, but k-means clustering can still work in the event of data without clear "borders" such as the iris dataset. Figure 9-8 shows the result of k-means clustering with three clusters, using petal widths and lengths as the input.

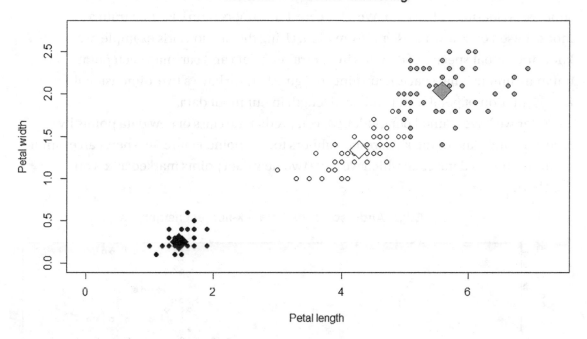

Iris Data -- k-means clustering

Figure 9-8. *Clusters of iris petals*

Looking carefully at Figure 9-8 vs. Figure 9-1, we can see that these cluster results do not match perfectly with the actual values of our data. The reason we end up in this situation is that at the margins between clusters, it can be hard to differentiate one type of iris from another. Therefore, the predictions might not be completely accurate.

If we think about how we might apply this process to outlier detection, we could calculate the centroids for each cluster and then, for each point, determine how far away it is from its cluster centroid. Data points that are relatively further away from other points of the same cluster are more likely to be anomalous. That said, we will not use k-means clustering as part of our multivariate outlier detection engine. The reason for this is that there are other, considerably more effective techniques for tracking cluster outliers available to us in the literature. What k-means clustering provides us is an easy-to-understand example of clustering. It also gives us a good introduction to the foundation for our multivariate outlier detection engine: k-nearest neighbors.

K-Nearest Neighbors

Unlike k-means clustering, the *k-nearest neighbors* algorithm is an example of a supervised learning algorithm. With k-nearest neighbors (knn), we first train a model based on known cluster information. Using the previous iris example, we know the actual species of each iris in our training data and can map each point out in n-dimensional space. Returning to Figure 9-1, we have a two-dimensional representation of petal width and petal length in our input data.

After we have trained our model, we can predict the class of new data points by reviewing the class of the k-nearest neighbors to that point. Figure 9-9 shows an example of this in the iris dataset, in which we have two new data points marked as x's on the chart.

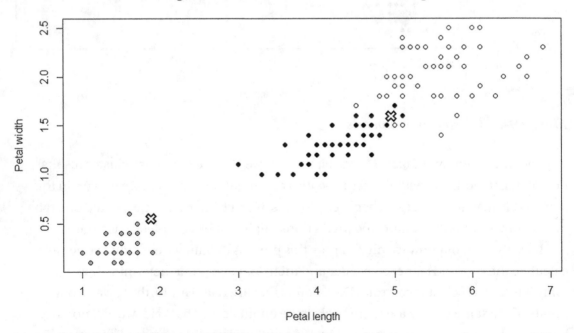

Figure 9-9. *Iris data with two new, unclassified data points*

Suppose that k=5. In that case, we want to find the class of the 5 nearest data points in order to find out to which class these new data points are most likely to belong. In the case of the first new data point, which we can see near the bottom left of the diagram, we can see that the five nearest data points all belong to the gray cluster. In this case, we can intimate that the new data point should belong to that gray cluster as well.

The second data point is trickier, as three of the five nearest data points are black and two of the five are white. This should give us some idea of the uncertainty involved in selecting an answer; that said, because more black than white data points are neighbors to our test point, we declare it to be a black data point as well. This leads to Figure 9-10, in which we update the colors of these new data points.

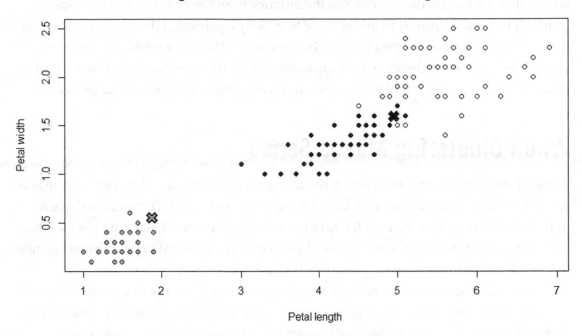

Edgar Anderson's Iris Data -- k-nearest neighbors

Figure 9-10. *The results of k-nearest neighbors classification*

This leads to an important note with knn: we want to choose odd numbers of neighbors, as doing so will give us a definite answer. With an even number, we might not end up with a concrete answer.

We can use two additional measures to gain an understanding of how confident the algorithm is in selecting the class: the number of neighbors sharing the same class and the average distance from those neighbors. In the case of the bottom-left example, all neighbors share the same class, and the new point is reasonably close to the cluster, indicating that this is very likely a member of the chosen class. In the second example, the neighbors are all reasonably close, but only 60% of them share the same class. This is a solid indication that the prediction could be wrong.

As we mentioned at the beginning of this section, knn forms the foundation of clustering outlier detection. It is not in itself a viable solution, however, as the k-nearest neighbors algorithm typically requires labeled data. This means that we would need to know if something was an anomaly before we trained a model. If you do have good information on whether specific data points were indeed anomalous, then knn can be a reasonable choice for an algorithm. In most cases, however, we won't have that amount of information and will need to discover outliers rather than identifying them.

Even so, the idea of proximity to neighbors plays a strong role in clustering-based outlier detection algorithms, as we use the distance from the centroid and the distance from our nearest neighbors as measures of how likely a particular data point is to belong to a given cluster. Over the next several chapters, we will look at algorithms that take these details into consideration. In the meantime, let's return to theoretical space and get an idea of when we might want to use clustering rather than some other technique.

When Clustering Makes Sense

We will look at two scenarios in which clustering algorithms make a lot of sense. The first scenario is multidimensional data, which is the theme of Part III. The intuition here is that combinations of variables will form in a variety of domains. So far, we have used the iris example to good effect, showing the physical characteristics of three different species of iris. We can also see clustering behavior between variables when a process is in different states, such as the combination of the current temperature of a gas range and its fuel utilization rate: when the range is on and hot, its utilization rate will be considerably higher than when it is off or when the burners are set to run at a lower temperature.

The other scenario of importance to us is multi-peaked data, which is what we ran into at the end of Chapter 8. Using a simple dataset like { 1, 2, 3, 50, 97, 98, 99 }, we can imagine that with more and more records coming in, we might see two separate peaks in the density of our dataset: one near the value 2 and one near the value 98. If our outlier detection process can understand that there are two separate groupings, we can discern that 50 is not really a member of either of these groupings. This is where a technique known as Gaussian Mixture Modeling comes into play.

Gaussian Mixture Modeling

Gaussian Mixture Modeling (GMM) is known as a *soft clustering* technique, meaning that for some observation, it assigns probabilities that the observation is in any given cluster. This is in contrast to *hard clustering* techniques, in which we must assign an observation to a single cluster. GMM is a probabilistic technique based on an attempt to discern *conditional distributions* from a known *joint distribution*. In other words, we can plot the density of our data points, as in Figure 9-11. This plot includes 33 points drawn from a normal distribution with a mean of 2 and standard deviation of 1, 40 points drawn from a normal distribution with a mean of 98 and a standard deviation of 1, and the number 50.0.

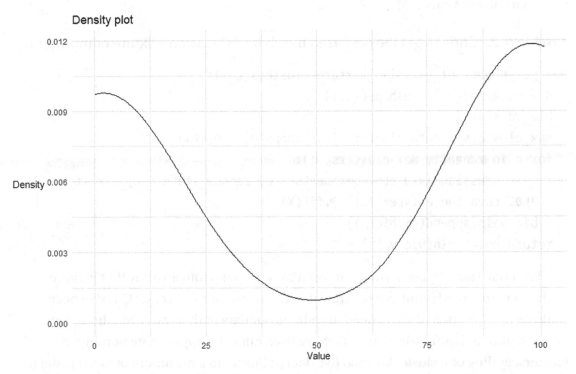

Figure 9-11. *A density plot exhibiting multi-peaked data. Note that this figure cuts off at 0 and 100; if we were to extend it, we would see the curves continue down and create two bell curves*

From this plot, we can clearly see the two peaks in our dataset, one at each end. The joint distribution of our data is clearly non-normally distributed. That said, we could imagine two conditional distributions that are normally distributed, one supporting each peak. GMM allows us to do just that for our univariate data. This will allow us to handle situations in which we may have multi-peaked univariate data before we move on to creating a proper multivariate clustering solution starting in Chapter 10.

Implementing a Univariate Version

Gaussian Mixture Models are available in scikit-learn, specifically `sklearn.mixture.GaussianMixture`. The process will include two major steps. The first step will determine the appropriate number of Gaussian Mixture clusters, as we will continue the practice of not assuming users have any foreknowledge of the distribution of their data. Listing 9-2 shows the code to determine the best number of Gaussian Mixture *components* (which we can think of as clusters).

Listing 9-2. Code to get the preferred number of Gaussian Mixture components

```
def get_number_of_gaussian_mixture_clusters(col):
  X = np.array(col).reshape(-1,1)
  bic_vals = []
  max_clusters = math.floor(min(col.shape[0]/5.0, 9))
  for c in range(1, max_clusters, 1):
    gm = GaussianMixture(n_components = c, random_state = 0, max_iter =
    250, covariance_type='full').fit(X)
    bic_vals.append(gm.bic(X))
  return np.argmin(bic_vals) + 1
```

The code starts by converting our DataFrame column into an array that includes each data individual point as a separate list. The reason for this is that GMM expects multivariate inputs, and so we need to reshape our data in that way, even though we actually have univariate data. After that, we determine the appropriate number of clusters, adding one cluster for each five data points up to a maximum of nine clusters. Admittedly, nine clusters is an artificial ceiling, but this should not be a problem for most practical datasets and does ensure that the operation finishes in a reasonable amount of time. Further, we should note that the calling code requires a minimum of 15 data points, meaning that the lowest value for `max_clusters` in practice will be 3.

Then, we run the `GaussianMixture` function for each number of components from 1 to the maximum number we calculate. Each execution has its own result metrics including one called the *Bayesian information criterion (BIC)*. This measure allows us to compare between GMM models, and we use np.argmin() to get the array element whose BIC has the lowest value, as lower values are preferable when comparing models. Because Python arrays start at 0 and we iterate sequentially from 1 component, we need to add 1 to the appropriate array element to get the best number of components for our given dataset.

Once we have the appropriate number of components, we will need to create a check to determine if we have any outliers based on our Gaussian Mixture Model results. Listing 9-3 includes the first half of the function for running a GMM check.

Listing 9-3. The beginning of the function to check for outliers via Gaussian Mixture Modeling

```
def check_gaussian_mixture(col, best_fit_cluster_count):
  X = np.array(col).reshape(-1,1)
  gm_model = GaussianMixture(n_components = best_fit_cluster_count, random_
  state = 0, max_iter = 250, covariance_type='full').fit(X)
  xdf = pd.DataFrame(X, columns={"value"})
  xdf["grp"] = list(gm_model.predict(X))
  min_num_items = math.ceil(xdf.shape[0] * .05)
  small_groups = xdf.groupby('grp').count().reset_index().query('value <= @
  min_num_items')
  small_groups["small_cluster"] = 1.0
  xdf = xdf.merge(small_groups[['grp', 'small_cluster']], on='grp',
  how='left').fillna(0)
```

This function starts with another run of the GaussianMixture() function. We could potentially revise the code in Listing 9-2 to return a GMM model, but for the sake of making it easier to test each component individually, I did introduce a bit of redundancy in the code. After getting the model, we break X into a new DataFrame called xdf. Then, we make a prediction using the predict() function. This returns the component number with the highest probability, as that is the best guess. There are two major checks we perform to see if a particular data point appears to be an outlier. The first check is to see if it belongs to a sufficiently small cluster. Our definition of "small" includes two aspects: first, if a cluster is less than 5% of the total number of data points, we consider it sufficiently small as to consist of outliers. Second, if a cluster includes only one data point, we definitely consider it to be an outlier—this is why the groupby() operation's query() function looks for values less than or equal to the minimum number of items. Even in cases with fewer than 20 data points, the call to math.ceil()

will guarantee that we capture solo data points. We capture the component numbers of all small clusters and assign an outlier score of 1.0 to all elements in xdf that belong to those small group components.

From there, we introduce the second part of outlier detection with GMM. One of the assumptions of GMM is that each individual component follows from a normal distribution—that's the *Gaussian* part of Gaussian Mixture Models. Therefore, all of our normal outlier detection techniques are available within the scope of each component. We could conceivably run all of the tests on individual components like we do entire datasets, but for the sake of parsimony, I include one test: if a data point is more than 3 MAD from the component's median, then we consider it an outlier. Listing 9-4 covers this part of the GMM check.

Listing 9-4. Running a MAD check per component and determining an outlier score based on the results of this check

```
for g in xdf["grp"].unique():
  xdf_g = xdf[xdf["grp"] == g]
  calc = perform_statistical_calculations(xdf_g["value"])
  xdf_g = xdf_g.drop_duplicates()
  if calc["mad"] > 0.0:
    xdf_g['far_off'] = [check_mad(val, calc["median"], calc["mad"], 3.0)
    for val in xdf_g['value']]
  else:
    xdf_g['far_off'] = 0.0
  for r in range(len(xdf_g)):
    xdf.loc[xdf['value']==xdf_g.iloc[r,0], "far_off"]=xdf_g.iloc[r,4]
    return [max(sc, fo) for (sc, fo) in zip(xdf["small_cluster"], xdf["far_off"])]
```

For each unique GMM component, we get all of the elements associated with that component and perform our statistical calculations. Then, we remove duplicate values because this process is deterministic: no matter how many times we calculate the difference for the value 1.5, it will always be the same given the calculated statistics in calc. Next, we check to see if MAD is greater than 0. If MAD is 0, that means all of our data points for this component contain the same value and so our measure of distance from MAD is 0. Assuming there is some variance in data points, we run the check_mad() function just as we do for the dataset as a whole. After performing these checks, we update xdf and set the value for far_off.

The last thing we do is return the larger of our calculation for small clusters (`small_cluster`) or distance from MAD (`far_off`). This gives us a score in approximately the same scale as our other scores and allows us to incorporate this new check.

In order to incorporate the check, we will need to make several updates. The first update is to the `weights` dictionary in the `detect_univariate_statistical()` function. Choosing a good weight is tricky and necessitates a trade-off. The reason for this is that things that will fail the GMM check may not fail any other check—in fact, they may look like ideal data points, as in our situation with a single value of 50 nestled between two equally sized clusters. Therefore, we want to ensure that the GMM weight is sufficiently high that it can trigger an outlier indication on its own. Increasing the sensitivity of this test means that it is more likely to have marginal cases fire as well. To balance these considerations, I landed on a weight of 1.5, although that might require further investigation with your datasets to ensure it is a sane value for you to use.

When it comes to running the test itself, we run GMM checks in the `run_tests()` function if and only if we have at least 15 data points in our dataset. GMM tends to perform better as datasets get larger and we can more easily pick out the individual components; with a small dataset, it can be difficult to figure out if some data point in fact belongs to a new component. Furthermore, if the function that returns the number of Gaussian Mixture components indicates that we have just one component, we can rely on our normal battery of checks.

One final change we will make is to the `determine_outliers()` function. Previously, we would return that a data point is an outlier if its anomaly score is strictly greater than the sensitivity cutoff—that is, the greater of the sensitivity score or the max fraction of anomalies score, assuming max fraction of anomalies is less than 1.0. Now, we will change this function to return that a data point is an outlier if its anomaly score is greater than or equal to the sensitivity cutoff. The reason for this is that looking only at data points whose anomaly score is greater than our cutoff could eliminate data points we consider anomalies, particularly when we have a large number of values with the same anomaly score. For example, suppose we have 100 data points and a max fraction of anomalies of 0.1. This means that we look for the anomaly score at the 90th percentile that, for the sake of discussion, we will say is 0.80. If 0.80 is higher than the sensitivity level, we will use this 0.80 as our sensitivity cutoff. Now suppose that the top 20 data points all have anomaly scores of 0.80. In this case, our original code would say that there are no anomalies because none of the data points have anomaly scores *greater than* 0.80—they are all equal to 0.80! With our new code, we will return 20 outliers due to ties.

This is likely to be more in line with what users would desire, though it does change the results of certain unit tests, particularly those tests in which the max fraction of anomalies is the limiting factor.

Updating Tests

Whenever we add a new check to an ensemble, there are likely to be changes at the margin. Introducing GMM is no exception, as there are several tests whose results change due to this introduction. The code repo accompanying this book indicates which tests specifically change, but a high-level summary is that we will often end up with more outliers as a result of introducing GMM.

We also add one new test case: `test_detect_univariate_statistical_multi_cluster()`. This test case runs tests whose input data consists of varying numbers of clusters. Its purpose is to ensure that we can understand how GMM interacts with our other algorithms and how our univariate outlier detection engine decides what is an inlier vs. an outlier. Figure 9-12 shows the visual interpretation of one test containing 30 observations. Twenty-four of these data points are numbered 1, 2, or 3. In addition, we have two 98s, two 99s, and one with a value of 100. Finally, we have our data point of 50.

Anomaly score per data point

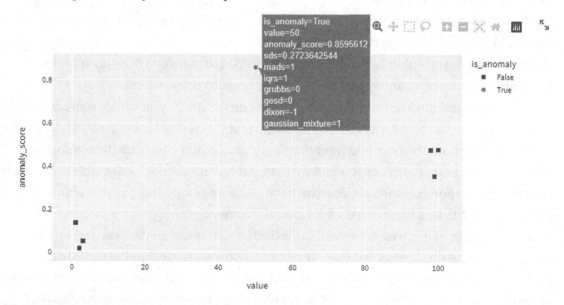

Figure 9-12. *50 is clearly an outlier*

Things become more interesting once we allow for some variance between values. Figure 9-13 shows a different scenario. In this scenario, we still have a large cluster of values ranging from approximately 1 to 3, a smaller cluster consisting of five data points ranging from approximately 98 to 100, and two additional values: one at 50 and one at -50. Our expectation is that we would have two outliers: 50 and -50.

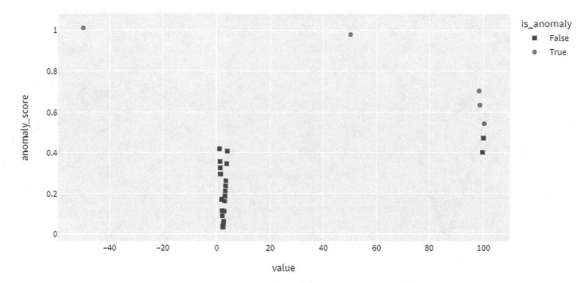

Figure 9-13. *Three elements from the larger cluster get pulled in as outliers*

In actuality, we have three additional values that end up being classified as outliers when we set the sensitivity score to 50. For these three additional data points, the Gaussian Mixture component of their weighted anomaly scores is fairly high, ranging between 0.22 and 0.48. This alone is not enough to get them marked as outliers. What pushes them over the edge is that the classic measures like MAD and interquartile range are high for these data points. The reason why is that the median for this dataset is close to 0 and there are more values in the 1–3 range than outside of that range. Because of this, values outside of 1–3 get a major penalty. The final two data points are close to being marked as outliers, but because they are closer to the center of the rightmost cluster, their Gaussian Mixture score components are not quite as high, and therefore, they escape being marked as outliers. If we do not want to see the three rightmost outliers, we might try various methods of tweaking our results, such as using a lower sensitivity level or modifying the weights for each component. No matter the technique, however, there will always be some number of marginal outliers that are not correctly classified.

On the other side of things, we can also see an accidental outcome of introducing Gaussian Mixture Modeling: once there are enough examples of outliers, they form their own cluster and are no longer truly outliers. Figure 9-14 shows an example of this, in which most data points range from 8.67 to 36.6 but there is then a cluster of values over 100.

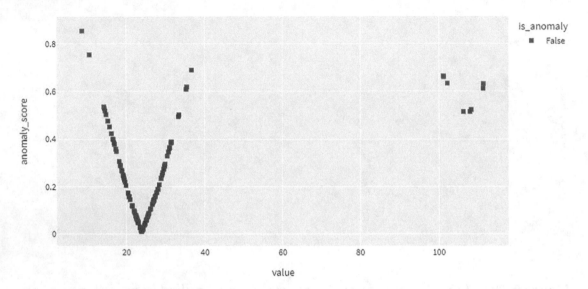

Figure 9-14. *A cluster of large values triggers high anomaly scores but not enough to be obvious outliers*

Without Gaussian Mixture Modeling, the other techniques are sufficient to point out these high values as outliers. With GMM, it turns out that there are enough values that they form their own cluster containing more than 5% of the total data. Because they hit the 5% threshold, that pulls their anomaly scores down just enough to escape without being marked as outliers. Depending on how sensitive we set our detection engine, we will still capture these high-value data points; along with them, however, we will also capture a fair number of values from the larger group that happen to have higher anomaly scores. At the end of the day, humans will still need to assess the results from any outlier detection engine.

Common Problems with Clusters

There are two important considerations when dealing with any clustering approach: choosing the right number of clusters and dealing with the inherent nondeterministic behavior of clusters.

Choosing the Correct Number of Clusters

Selecting the correct number of clusters can be a challenge, especially when we introduce the expectation that users do not have a full awareness of the shape of their data. In the Gaussian Mixture Modeling code shown previously, we saw one potential solution to this problem: constrained maximization. In the GMM case, we minimized (remembering that minimization is just maximization in the opposite direction!) the Bayesian Information Criterion (BIC) score for models with differing numbers of components. BIC is particularly useful because it includes a penalty for complexity, meaning that the score biases toward a solution having fewer components if it is approximately as good as one with more components. Therefore, adding another component requires a significant improvement, especially as the total number of components grows relatively large.

Another approach is common with algorithms like k-means clustering: the "elbow method." The elbow method uses within-cluster sum of squares (WCSS). Importantly, the goal here is not to minimize the WCSS, as it would be trivial to do so: have one cluster for every data point. That solution completely defeats the purpose of clustering, however, and so we need to have a better answer in mind. In this case, the elbow method gets its name because we are looking for changes in WCSS as we introduce more clusters. Figure 9-15 shows an example of an elbow plot using the iris dataset.

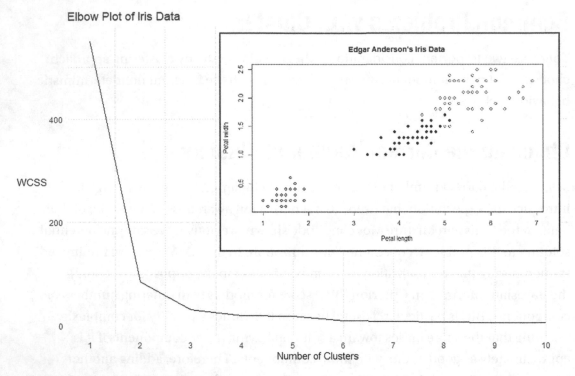

Figure 9-15. *An elbow plot showing within-cluster sum of squares changes as the number of clusters increases. An inset of the initial plot is included for ease of comparison*

We can see a dramatic difference in moving from one cluster to two, and a brief review of the original data helps us understand why: there are two major clumps of data. In moving from two to three clusters, we still see an improvement though not as drastic of one. Beyond three clusters, we see a minimal improvement in WCSS. Therefore, we conclude that either two or three clusters would be the best for this dataset.

These are not the only techniques we can use for determining the appropriate number of clusters, but they are two techniques that do not require any domain knowledge.

Clustering Is Nondeterministic

The other problem we might run into with clustering is that the results are not guaranteed to be the same every time. Some clustering algorithms, including k-means clustering, are nondeterministic. Different initial positions of each centroid can lead to different results. Figure 9-16 shows an example of this in action. One thing to note is that in order to get the results from Figure 9-16, I did need to include two additional variables that are not graphed: sepal length and sepal width.

Iris Data -- k-means clustering

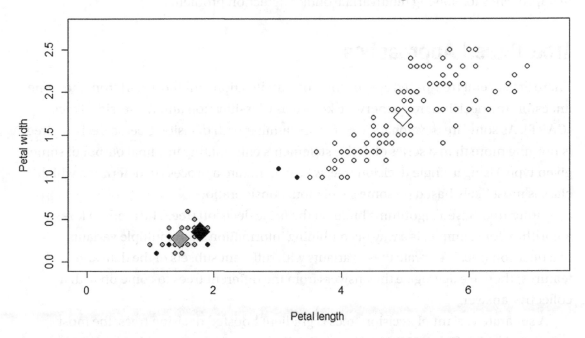

Figure 9-16. *Multiple solutions to the same clustering problem are possible. In this case, we have additional features that we do not display, leading sometimes to this uneven result and sometimes to the result we see in Figure 9-1*

The additional features are not a necessary condition for cluster variability but they do provide us a concrete, replicable example. This can be a bigger issue for testing as we introduce more features. When it comes to outlier detection, cluster instability may affect our results at the margins, though not as badly as Figure 9-16 would indicate. The reason is that Figure 9-16 only looks at two of the four relevant variables; points which might appear to be outliers based solely on these two variables may in fact be much closer to the centroid when taking all relevant features into consideration.

Alternative Approaches

Our focus in this part of the book is clustering approaches but we would be remiss in ignoring altogether other viable approaches. In this section, we will cover another class of approaches for solving multivariate outlier detection problems: trees.

Tree-Based Approaches

There are several tree-based approaches for classification, but most start from the same ancestor: the decision tree, otherwise known as Classification and Regression Trees (CART). As software developers, we are very familiar with decision trees: a decision tree is nothing more than a series of if-else statements culminating in a final output of some given type. Using a single decision tree, we can generate a process to determine which class is most likely based on some set of input considerations.

Other tree-based algorithms build on the basic decision tree. The random forest algorithm, for example, is a way of combining information from multiple variants of a decision tree. We create these variants with different subsets of the data and features; then, we aggregate the answers from the different trees to come up with a collective answer.

A separate variant of decision trees is gradient boosted decision trees, the most popular of which is the XGBoost library. Gradient boosting works with decision trees sequentially, meaning that the algorithm takes information learned from prior models and uses that information to create a better model. It does so by determining what the prior model got wrong and adds a modification factor intended to choose the correct answer more often than in the prior iteration.

The isolation forest algorithm is an example of a tree-based approach that is not rooted in decision trees. Instead of assigning a given observation to a class based on conditional rules, the isolation forest algorithm randomly partitions the data and figures out how many of these random partitions it takes to isolate a given data point. The fewer random partitions it takes, the more likely it is that a given data point will be an outlier, as it implies that the data point is not particularly close to other data points.

The Problem with Trees

Tree-based algorithms are great for classification, especially algorithms like XGBoost. When it comes to outlier detection, however, tree-based algorithms are not as good. Tree-based methods will tell you the class to which an item belongs but typically will not give you any kind of likelihood score. Without a likelihood score, we need to revert to asking if a given data point is an outlier and answering "yes" or "no" to the question but without the additional context that we might get from a distance-based approach. Furthermore, when we do reduce outlier detection to a classification problem, we now create an issue with imbalanced classes: outliers, by definition, are rare—if they were common, they'd be inliers! Depending on how rare outliers are in a given dataset, a classification algorithm might not be able to detect outliers at all, as classification algorithms tend to have difficulty working with imbalanced class data. In addition to these issues, training a classification algorithm means that we need to have labeled data on which elements were outliers and which were not. Most of the time, we will not have a clean dataset with this labeled information.

Isolation forests attempt to mitigate this first issue of limited information by way of calculating a measure of distance from other points in the data. This does help resolve the first issue with a classification-based tree, so that is a point in favor of isolation forests. The biggest downside is that the isolation forest algorithm is often not as accurate as the approaches we will deal with in Chapters 10–12. Therefore, we will not include isolation forests in this book.

Conclusion

This chapter marks a turning point in the book. The clustering techniques we have covered, particularly Gaussian Mixture Models, were useful in filling a gap in our univariate capabilities. As we close the door on univariate anomaly detection, this clustering analysis gives us insight that will be helpful throughout the rest of Part III.

CHAPTER 10

Connectivity-Based Outlier Factor (COF)

The prior chapter provided us with an introduction to clustering techniques and the use of one such clustering technique, Gaussian Mixture Models. Over the course of this chapter, we will implement a separate technique for multivariate clustering: Connectivity-Based Outlier Factor.

Distance or Density?

There are two key approaches to outlier detection: distance-based approaches and density-based approaches. For univariate anomaly detection, we generally focused on distance-based approaches. As Mehrotra explains, "A large distance from the center of the distribution implies that the probability of observing such a data point is very small. Since there is only a low probability of observing such a point (drawn from that distribution), a data point at a large distance from the center is considered to be an anomaly" (Mehrotra, 97). We started off by implicitly assuming that our univariate datasets fit one cluster, a cluster whose shape might approximate a normal distribution. By Chapter 9, we relaxed that assumption and allowed for multiple clusters using Gaussian Mixture Models. In doing so, we also introduced a density-based approach to detecting outliers.

Figure 10-1 provides an example of the importance of density-based approaches for outlier detection. If we use distance as the sole measure of outlier-ness, then the marked data point in the left-hand cluster will have a higher anomaly score than the marked data point in the right-hand cluster. Our gestalt sensibilities, however, disagree with such an assessment: we can see that although vector A is in fact longer than vector B, the data point associated with vector A fits in a regular pattern with other data points. Each data point is roughly equidistant from at least two neighbors. By contrast, the data point

185

© Kevin Feasel 2022
K. Feasel, *Finding Ghosts in Your Data*, https://doi.org/10.1007/978-1-4842-8870-2_10

associated with vector B is further away from a neighbor than any of its compatriots. In other words, for density-based approaches, we care about the distance between neighbors more than the distance from the centroid.

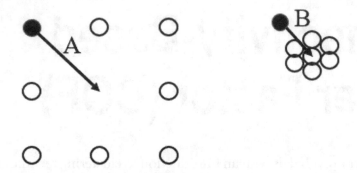

Figure 10-1. *Vector A is longer than vector B. Therefore, a pure distance-based approach may lead us to believe that point A is more of an outlier than point B*

With density-based approaches, we think in terms of neighborhoods, which we can define as a given number of nearest neighbors. We saw this in practice in Chapter 9 with the k-nearest neighbors (knn) algorithm. What differentiates knn from the density-based outlier detection algorithms we will use in subsequent chapters is that knn stops after defining the neighborhood. With density-based outlier detection algorithms, we go one step further: after determining what the neighborhood is for a given point, we then compare the density of that data point vs. its neighbors. If the data point's density is considerably smaller than that of its nearest neighbors, then the data point is likely to be an anomaly.

Figure 10-2 shows an example of this concept. In this case, we show an example of points along a single dimension. Note, however, that the multidimensional version of this problem is conceptually the same, except that we need to use some distance measurement like *Manhattan distance* or *Mahalanobis distance* to calculate the distance between points. These measurement techniques are outside the scope of this book but are of some relevance when digging into algorithms.

Figure 10-2. *In the top image, we show the distance from our data point (in black) to its two nearest neighbors. In the bottom image, we show the distance from each of those neighbors to their two nearest neighbors*

Reviewing the results in Figure 10-2, we can see that the average distance between points is considerably higher for our data point vs. its neighbors. We define the *local density* of a point as the reciprocal of that average distance between points. Therefore, our data point is in a lower-density area than its neighbors, and this is an indicator of a data point being an outlier.

One important rule that has to hold to make this outcome possible is that the set of neighbors is not *reflexive*. In other words, if we look at the two nearest neighbors to the black point, neither of them has the black point as one of its two nearest neighbors. As a rule of thumb, the fewer the number of neighbors a data point has that call it a nearest neighbor, the more likely that data point is to be an outlier.

Before we can get to the Connectivity-Based Outlier Factor approach to clustering, we need to understand another algorithm first.

Local Outlier Factor

The *Local Outlier Factor* (LOF) algorithm is a density-based approach to finding outliers in a dataset. We measure how the local density around a data point compares to that of its k nearest neighbors and report a data point as an outlier if its LOF is sufficiently large compared to others in the same cluster. Suppose we have some data point x and a neighborhood size of k=5. We can then find the distance between x and its 5th neighbor, as we see in Figure 10-3. We can then visualize this as a circle around x, intersecting the kth point from x.

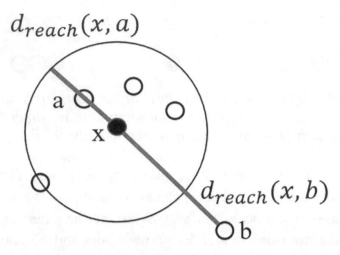

Figure 10-3. *Calculating reachability distance for data point x with k=5 with respect to two other data points a and b. Note that the neighborhood size includes data point x*

We then calculate the distance from x to each point, taking the larger of the actual distance or the distance to the kth point as d_{reach}. For example, point a lies within the boundaries of our circle, and so we use the distance to the kth point to define the *reachability distance* from x to a. Point b, meanwhile, lies beyond our kth point, and so we use the actual distance from x to b as the reachability distance from x to b. After performing this operation for each pairing of data points in our extended neighborhood, we can calculate the average reachability distance for x by summing up all of the reachability distances from x to various points and dividing by the number of points in the cluster minus one (i.e., the denominator will not include data point x itself).

The average reachability distance gives us an idea of how far we need to travel in order to find similar data points. If we take the reciprocal of this, we now calculate the *reachability density* of a point: How packed-together are other data points near our point x?

Finally, we define the Local Outlier Factor as a ratio of the local reachability densities of all other data points vs. the local reachability density of point x. A relatively high value for this measure means that there tends to be a larger distance between data point x and its neighbors compared to other data points in the cluster. This result comports with one an intuitive understanding of outliers: data points that are far removed from others are more likely to be outliers than ones packed closely together. Once we have the LOF per data point, we can use that score to determine whether a particular data point is an outlier. One way we could do this is to take the top few percent of values and declare those data points as outliers. Alternatively, we could compare the largest scores and use techniques like MAD to see if there is a significant enough spread from the median to declare any given point a likely outlier.

Connectivity-Based Outlier Factor

The Local Outlier Factor algorithm can be quite useful for finding outliers, but it does have some practical limitations. The biggest limitation is that outliers do not always need to be in lower-density areas. For example, Tang et al. devise the following cluster that we see in Figure 10-4.

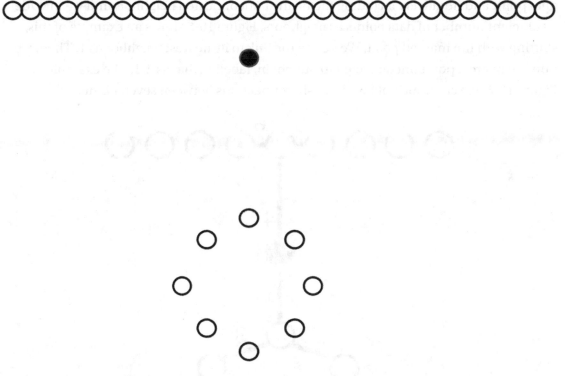

Figure 10-4. *Any value of k that allows LOF to declare the highlighted point as an outlier will also declare all points in the circle as outliers*

Here, we can see the outlier, which I have marked as a black dot for the sake of clarity. Note further that there is really only one outlier in this figure: we have a mass of points at the top of the image and a second cluster of points with some separation between them. Using the LOF algorithm, however, we run into a problem: in order to detect the marked point as an outlier, we would also classify all of the points in the circle as outliers because of the difference in distances between elements in the two clusters.

Connectivity-Based Outlier Factor (COF) is another density-based approach to clustering, one that attempts to resolve these sorts of problems with LOF. It does so by adding to density an idea called *isolativity*: the degree to which an object is connected to other objects. The points making up the circle are connected to one another, just as the points in the line at the top of the figure are connected to one another. The marked point, however, does not fit either of the established patterns—it is isolated from those patterns and is therefore more likely to be an outlier than a point that is connected in a pattern.

Following is a high-level summary of how COF works. For a given data point and the number of neighbors we care about, we will build a chain, starting with our given data point and iteratively including the nearest unconnected neighbor until we include a sufficient number of data points as neighbors. Figure 10-5 shows an example of this, starting with the marked point. We define the link to its nearest neighbor as 1. Then, we find the nearest point unconnected to our chain, labeling that as 2. In the example of Figure 10-5, we continue until we have six connections between seven points.

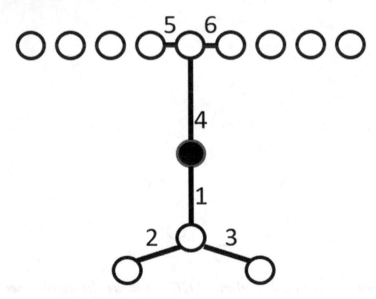

Figure 10-5. *The six connections for the marked data point. Note that edges are enumerated sequentially in order of the shortest path from the already-connected chain to the next free point*

Once we have these connections, we perform a weighted calculation, emphasizing the "earlier" (numerically lower) connections more than the "later" (numerically higher). This provides us the Connectivity-Based Outlier Factor (COF) value. Data points with higher COF values are more likely to be outliers than data points with lower COF values. This allows us to sort data points by COF value in descending order and find the biggest outliers, something that is particularly handy if the base algorithm returns more outliers than our maximum fraction of anomalies parameter allows.

The math for calculating COF is available in Tang et al., but because we are going to use the Python Outlier Detection (PyOD) library to calculate COF, we will not include it here. Instead, we will move on to implementation.

Introducing Multivariate Support

The code for multivariate anomaly detection will necessarily look somewhat different from univariate anomaly detection—after all, the data structure for multivariate anomaly detection includes a key and *list* of values vs. a key and a single value for univariate anomaly detection. Even so, there will be considerable overlap in the groundwork code.

Laying the Groundwork

Listing 10-1 shows the API endpoint definition for multivariate anomaly detection. In it, our detection method includes four parameters relevant to detection. Two of them, sensitivity score and max fraction of anomalies, match the univariate example. A third, input data, is functionally similar to the univariate scenario, although instead of having a single value `val`, we have a list of values `vals`. Each element in the list represents a different feature we care about. For example, if we wish to perform outlier detection on a set of prospective professional athletes, we might include attributes such as height, weight, straight-line running speed, agility tests, and so on. Each of these would be a separate value in our list, and we would compare prospective athletes across the entire set of measures, not just one. The fourth parameter is new: `n_neighbors`, which represents the desired number of neighbors. This parameter can make a considerable difference in whether we label something as an outlier, but there is limited general advice on selecting the best number. Typically, the advice is to perform a search along a range from 10 (assuming you have at least 30 data points) to as high as 100 or 125 (assuming a sufficiently large number of data points). For the sake of development, we will leave this as an option for the caller for now, though we will come back to this decision in the next chapter.

Listing 10-1. Offering multivariate outlier detection to outside callers

```
class Multivariate_Input(BaseModel):
  key: str
  vals: list = []

@app.post("/detect/multivariate")
def post_multivariate(
  input_data: List[Multivariate_Input],
  sensitivity_score: float = 50,
  max_fraction_anomalies: float = 1.0,
  n_neighbors: int = 10,
  debug: bool = False
):
  df = pd.DataFrame(i.__dict__ for i in input_data)

  (df, weights, details) = multivariate.detect_multivariate_statistical
  (df, sensitivity_score, max_fraction_anomalies, n_neighbors)

  results = { "anomalies": json.loads(df.to_json(orient='records')) }

  if (debug):
    results.update({ "debug_weights": weights })
    results.update({ "debug_details": details })
  return results
```

Based on this definition, we know that we will need a detect_multivariate_statistical() function in our code base. We will also want helper functions like run_tests() and determine_outliers(), similar to our univariate example.

Listing 10-2 shows the implementation of detect_multivariate_statistical(). Just as in the univariate scenario, we will weigh algorithms differently and therefore need a dictionary of weights. We determine the number of data points in the dataset and then perform some basic checks (removed from the following listing but available in the code accompanying this book at https://github.com/Apress/finding-ghosts-in-your-data), ensuring that we have a sufficient number of data points to perform multivariate outlier detection and that the inputs are reasonable. In addition, because our implementation of COF requires that no more than 50% of data points be anomalies, we silently cap max fraction of anomalies at 0.5 to ensure this.

Listing 10-2. A function to detect outliers in multivariate data. Certain diagnostic functionality has been removed from this listing for the sake of brevity, but it is available in the accompanying repository.

```
def detect_multivariate_statistical(
  df,
  sensitivity_score,
  max_fraction_anomalies,
  n_neighbors
):
  weights = {"cof": 1.0}

  num_data_points = df['vals'].count()
  if max_fraction_anomalies > 0.5:
      max_fraction_anomalies = 0.5
  (df_encoded, diagnostics) = encode_string_data(df)
  (df_tested, tests_run, diagnostics) = run_tests(df_encoded, max_fraction_
  anomalies, n_neighbors)
  df_out = determine_outliers(df_tested, sensitivity_score, max_fraction_
  anomalies)
  return (df_out, weights, { … })
```

Once we have determined that the dataset and input parameters are all viable, we need to encode any string data that comes in. Outlier detection algorithms deal almost exclusively in numeric features, but our datasets often have strings that represent values. We could have the caller make this conversion, but for the sake of convenience and to support a wider variety of datasets, we will introduce an encode_string_data() function that translates any string column to an ordinal. Listing 10-3 shows this function. The first step of the function is to break out the list in vals and turn it into a set of columns. We don't know exactly what the columns represent, but the lack of meaningful names does not matter for our purposes. The next step is to determine whether we have any string columns using the select_dtypes() function in Pandas. If there are string columns, we create an ordinal encoder and translate each unique string into a corresponding unique number. We then merge together the two DataFrames, which we can do because they will contain the same number of rows and will be in the same order. This will ensure that we have the original data as well as numeric-only translations.

Listing 10-3. Encoding string data

```
def encode_string_data(df):
  df2 = pd.DataFrame([pd.Series(x) for x in df.vals])
  string_cols = df2.select_dtypes(include=[object]).columns.values
  diagnostics = { "Number of string columns in input": len(string_cols) }
  if (len(string_cols) > 0):
    diagnostics["Encoding Operation"] = "Encoding performed on string
    columns."
    enc = OrdinalEncoder()
    enc.fit(df2[string_cols])
    df2[string_cols] = enc.transform(df2[string_cols])
  else:
    diagnostics["Encoding Operation"] = "No encoding necessary because all
    columns are numeric."
  return (pd.concat([df, df2], axis=1), diagnostics)
```

Even though we do include an encoding function, it is important to note that the way we are encoding is easy but not necessarily a good practice. The reason is that our encoder is ordinal and does not include any measurement of string "nearness," either in terms of how many letters apart two words are or how similar (or dissimilar) the meanings of two words are. As an example, the word "cat" might have an ordinal value of 1 but "cats" a value of 900. Our transformation is simple and likely will not work well on text-heavy datasets. The intent here is to provide an easy method for working with a mostly numeric dataset that happens to have one or two categorical string features, rather than analyzing the textual content of features and discerning meaning from them. It is possible to create a more sophisticated encoder that does this, but that is an exercise for another book on natural language processing.

Implementing COF

The actual implementation of COF is simple. Listing 10-4 includes the run_tests() function as well as the check_cof() function it calls.

Listing 10-4. Implementing the check for Connectivity-Based Outlier Factor

```
def run_tests(df, max_fraction_anomalies, n_neighbors):
  tests_run = {
    "cof": 1
  }
  (df_cof, diagnostics) = check_cof(df, max_fraction_anomalies,
  n_neighbors)
  return (df_cof, tests_run, diagnostics)

def check_cof(df, max_fraction_anomalies, n_neighbors):
  col_array = df.drop(["key", "vals"], axis=1).to_numpy()
  clf = COF(n_neighbors=n_neighbors, contamination=max_fraction_anomalies)
  clf.fit(col_array)
  diagnostics = {
    "COF Contamination": clf.contamination,
    "COF Threshold": clf.threshold_
  }
  df["is_raw_anomaly"] = clf.labels_
  df["anomaly_score"] = clf.decision_scores_
  return (df, diagnostics)
```

The run_tests() function is very similar to its univariate counterpart, saving the interesting code for check_cof(). This latter function accepts our DataFrame as well as the max fraction of anomalies and desired number of neighbors. It first creates an array out of all columns except key and vals, as the COF function requires inputs be shaped as an array rather than a DataFrame. After creating a classifier function using COF(), the function fits the classifier to our input array. We use the corresponding labels_ attribute to see if COF considers a particular data point as anomalous and decision_scores_ to gain an idea of how much of an outlier a given data point is. The *contamination factor*, which we capture as clf.contamination, is an optional input that defaults to 0.1 and represents the proportion of outliers in the dataset. This is similar to what we call the max fraction of anomalies in our engine, except that contamination factor requires that we mark a certain percentage of results as outliers, no matter how high or low the decision score is. If we set the contamination factor to 0.5, half of our data points will be marked as outliers, even if every data point is tightly clustered together and a human would not recognize any data points as anomalies. The threshold_ output provides the cutoff point for what COF labels as an outlier.

The most important numeric attribute we get from COF is the decision score. This numeric value is what we send to the determine_outliers() function. Listing 10-5 shows the code for this function. The code is somewhat similar to the univariate scenario, but it also accounts for COF's determination of what makes an outlier.

Listing 10-5. The outlier determination function for multivariate outlier detection

```
def determine_outliers(
  df,
  sensitivity_score,
  max_fraction_anomalies
):
  second_largest = df['anomaly_score'].nlargest(2).iloc[1]
  sensitivity_score = (100 - sensitivity_score) * second_largest / 100.0
  max_fraction_anomaly_score = np.quantile(df['anomaly_score'], 1.0 - max_
fraction_anomalies)
  if max_fraction_anomaly_score > sensitivity_score and max_fraction_
anomalies < 1.0:
    sensitivity_score = max_fraction_anomaly_score
  return df.assign(is_anomaly=df['anomaly_score'] > np.max([sensitivity_
score, 1.35]))
```

We start by calculating the sensitivity score. Because our anomaly score range may be quite different from 0 to 1, we adjust the sensitivity_score range to run from 0 to the second-largest anomaly score, dealing with cases in which one extreme outlier dominates everything else. After doing that, the calculations are the same as the univariate scenario. An important note here is that, unlike the univariate scenario, we could always get back a number of anomalies equal to max_fraction_anomalies * number of data points when sensitivity score is set to 100. COF does not have any in-built rules around whether a particular point is an outlier or not, making our selection of max fraction of anomalies and sensitivity score even more important than in the univariate scenario. That said, values closer to 1.0 are less likely to be outliers, and choosing some minimum threshold will reduce the number of spurious outliers. 1.35 is an arbitrary choice but one that is intended to prevent an outcome in which a dataset with no actual anomalies triggers a large number of reported outliers.

Test and Website Updates

This new technique brings with it a new set of tests, as well as some challenges we will need to face regarding visualization of multivariate data. Let's start first with unit test changes, move on to integration tests, and end with website updates.

Unit Test Updates

The file test\test_multivariate.py contains the unit tests we will use to confirm functionality in our multivariate outlier detection scenario. Listing 10-6 covers the first test, which ensures that we encode strings as ordinals properly. To make it easier to read, only one test case appears in the listing; the remaining test cases are available in the test file.

Listing 10-6. Encode strings as ordinal values

```
@pytest.mark.parametrize("df_input, requires_encoding, number_of_string_
columns", [
  ([["s1a", [1, "Bob", 2, "Janice", 3]], ["s2", [4, "Jim", 5, "Alice", 6]],
  ["s3", [7, "Teddy", 8, "Mercedes", 9]]], True, 2),
)
def test_detect_multivariate_encoding_string_columns(df_input, requires_
encoding, number_of_string_columns):
  # Arrange
  df = pd.DataFrame(df_input, columns=["key", "vals"])
  # Act
  (df_encoded, diagnostics) = encode_string_data(df)
  encoding_performed = (diagnostics["Encoding Operation"] == "Encoding
  performed on string columns.")
  num_string_columns = diagnostics["Number of string columns in input"]
  # Assert
  assert(requires_encoding == encoding_performed)
  assert(number_of_string_columns == num_string_columns)
```

Reviewing the aforementioned, there are two string columns per array. The encoding function correctly sees this and dutifully encodes the strings in the second and fourth columns of the array.

Aside from this, there are several other functions intended to test COF's outputs and how sensitivity score and max fraction of anomalies combine to affect outcomes. There are three datasets that we will use in the tests. The first is *sample_input*, which contains an artificially generated set of data spanning four columns. It includes a fairly wide range of values, making it difficult to determine a priori whether there are any outliers. The second dataset is *sample_input_one_outlier*, which includes dozens of extremely similar rows, differing by no more than ~1%, as well as one radically different row. Finally, *sample_input_no_outliers* is the prior dataset minus the one outlier.

The minimum anomaly score check in `determine_outliers()` is critical for scenarios like *sample_input_no_outliers*. If we have no minimum threshold and set maximum fraction of anomalies to 1.0, we would need to drop the sensitivity score down to 8 in order to have fewer than ten outliers returned. With no cutoff and very close data points, instead of having no outliers, everything becomes an outlier.

Integration Test Updates

Along with new unit tests, we also need new integration tests to handle multivariate data. These tests are in the Chapter 10 Postman collection and handle fairly basic scenarios ranging from zero to three outliers in a small, 15-record sample. The most interesting of these tests is `Ch10 - Multivariate - Debug - Three Outliers, 2S1L - Two Recorded`. This test includes three outliers at three different orders of magnitude. The first outlier has an anomaly score over 100, the second has a score of 34, and the third has a score of 5.8. All three are legitimate outliers, but unless we set the sensitivity score to 83 or higher, we only capture the two larger outliers because the largest outlier "sets the curve" and makes it harder for us to find smaller outliers. This points at a flaw in our solution, one we aim to rectify over the next two chapters by introducing additional algorithms and forming a new type of ensemble.

Website Updates

The final section of this chapter will cover updates to our Streamlit dashboard. We first review the `process()` function in Listing 10-7. This listing includes all of the parameters we will need to send in for univariate or multivariate analysis. Note that we do not include the number of neighbors in this function, as this is an implementation detail that we do not want to force users to handle.

Listing 10-7. The definition for the `process()` function

```
@st.cache
def process(server_url, method, sensitivity_score, max_fraction_anomalies,
debug, input_data_set):
  full_server_url = f"{server_url}/{method}?sensitivity_score={sensitivity_
  score}&max_fraction_anomalies={max_fraction_anomalies}&debug={debug}"
  r = requests.post(
    full_server_url,
    data=input_data_set,
    headers={"Content-Type": "application/json"}
  )
  return r
```

The major change comes when selecting the *Detect!* button. We now need to break out the univariate and multivariate scenarios. Listing 10-8 shows this breakout, eliding over univariate code in the process.

Listing 10-8. Creating a visual to expose multivariate outlier results

```
if method=="univariate" and convert_to_json:
    input_data = convert_univariate_list_to_json(input_data)
  resp = process(server_url, method, sensitivity_score, max_fraction_
  anomalies, debug, input_data)
  res = json.loads(resp.content)
  df = pd.DataFrame(res['anomalies'])

  if method=="univariate":
    # ...
  elif method=="multivariate":
    st.header('Anomaly score per data point')
    colors = {True: '#481567', False: '#3CBB75'}
    df = df.sort_values(by=['anomaly_score'], ascending=False)
    g = px.bar(df, x=df["key"], y=df["anomaly_score"], color=df["is_
    anomaly"], color_discrete_map=colors,
```

```
        hover_data=["vals"], log_y=True)
st.plotly_chart(g, use_container_width=True)

tbl = df[['key', 'vals', 'anomaly_score', 'is_anomaly']]
st.write(tbl)

if debug:
  col11, col12 = st.columns(2)

  with col11:
    st.header('Debug weights')
    st.write(res['debug_weights'])

  with col12:
    st.header("Tests Run")
    st.write(res['debug_details']['Tests run'])
    st.write(res['debug_details']['Test diagnostics'])

  st.header("Full Debug Details")
  st.json(res['debug_details'])
```

With univariate data, we were able to create a scatter plot, using the one variable as the X axis and the anomaly score as the Y axis. With multivariate data, we are not able to plot the data directly. There are two reasons for this: first, we do not know how many dimensions there will be, making it difficult to discern what kinds of plots might even be useful here. Second, humans have a lot of difficulty making sense of three-dimensional plots, so even if we use a technique like Principal Component Analysis (PCA) to narrow our results down to two dimensions plus an anomaly score (assuming we are able successfully to get it down to two), displaying both values plus anomaly score in a way that humans can easily interpret becomes a challenge. Instead, we display each data point as a bar in a column chart, with the height of the bar representing the anomaly score. Because there can be a wide range in score values, the anomaly score is measured on a logarithmic axis. This allows us to see the differences in those smaller scores while still acknowledging the anomalous values. Figure 10-6 shows an example of this.

Anomaly score per data point

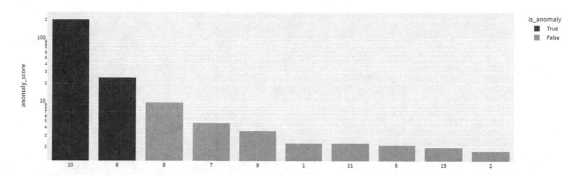

Figure 10-6. *The outcome of a multivariate anomaly detection test. This shows the key as the X axis and the anomaly score on a logarithmic scale as the Y axis*

After that, we display a table of all keys and values. Finally, if the user wishes to see debug information, we can see debug weights, tests run, and test diagnostics below the table.

Conclusion

In this chapter, we looked at one useful algorithm for multivariate outlier detection: Connectivity-Based Outlier Factor (COF). This is a good first algorithm to look at because it tends to be quite effective. As we saw in the test cases, though, understanding the appropriate cutoffs can be a challenge. In the next chapter, we will introduce another, complementary technique to COF and see how to combine the two together in an ensemble to reduce some of the challenge of picking appropriate inputs and cutoff values.

## CHAPTER 11

Local Correlation Integral (LOCI)

Chapter 10 introduced us to one multivariate clustering technique in Connectivity-Based Outlier Factor (COF). COF is a popular technique for multivariate outlier detection, but it does come with the downside that there is little immediate guidance outside of "higher outlier scores are worse." We lack solid guidance on what, exactly, represents a sufficiently large outlier score. In this chapter, we introduce a complementary technique to COF that does come with its own in-built guidance: Local Correlation Integral.

Local Correlation Integral

Local Correlation Integral (LOCI) is another density-based approach like Local Outlier Factor (LOF) and Connectivity-Based Outlier Factor (COF). Papadimitriou et al. describe one key difference between LOCI and other density-based measures: LOCI "provides an automatic, data-dictated cut-off to determine whether a point is an outlier" (Papadimitriou et al., 1). This built-in cutoff means that users will not need to provide any information outside of the data itself. Over the course of this section, we will see how the algorithm works; then, we will incorporate it into our multivariate outlier detection engine.

Discovering the Neighborhood

An important difference between LOCI and other density-based approaches like LOF or COF is that LOCI does not ask for an explicit number of neighbors in a neighborhood. Instead, it takes a measure called *alpha (α)*, which represents neighborhood size, with $0 < \alpha < 1$. The default value for α is 0.5; Papadimitriou et al. chose this value for their experiments (Papadimitriou et al., 5), and we will stick with this value.

Along with α, we need a variable r, which represents the distance of our *sampling neighborhood*: the set of points we will use against which to compare a given point. Specifically, we determine if that data point is an outlier based on the number of neighbors within a bounded circle measuring α * r, otherwise known as the *counting neighborhood*. Figure 11-1 provides a visual example of this.

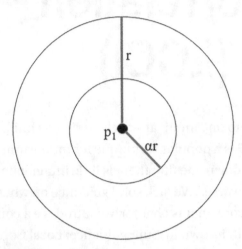

Figure 11-1. *A representation of the sampling neighborhood (with radius of r) and counting neighborhood* α * r

Within our counting neighborhood α * r, we count the number of data points that exist, giving us an idea of how densely packed data points are near our point of interest. Then, to gain an understanding of how that compares to other points, we repeat the process for each point inside our sampling neighborhood. Figure 11-2 fills in additional detail, introducing two additional points: p_2 and p_3. Because these two points exist inside p_1's sampling neighborhood, we use these two points to compare against p_1. The results we come up with are that p_1 has one data point in its counting neighborhood, p_2 has three, and p_3 has five, noting that counts are inclusive of the tested data point.

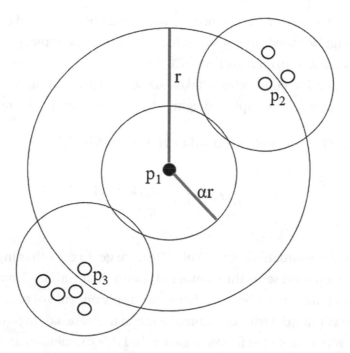

Figure 11-2. *The counting neighborhoods of three data points within p_1's sampling neighborhood*

From the image in Figure 11-2, we can determine two important criteria. The first is the number of data points inside the counting neighborhood, which we can represent as $n(p_i, \alpha r)$. This comes out to 1 for p_1, 3 for p_2, and 5 for p_3, respectively.

The second criterion we can determine is the mean of αr-neighbors within a given sampling neighborhood, which we can represent as $\hat{n}(p_i, r, \alpha)$. In this case, the mean is 3.

With these two criteria, we can calculate the *Multi-granularity Deviation Factor*.

Multi-granularity Deviation Factor (MDEF)

The Multi-granularity Deviation Factor (MDEF) is defined using the formula in Listing 11-1.

Listing 11-1. The formula for Multi-granularity Deviation Factor

$$MDEF(p_i, r, \alpha) = 1 - \frac{n(p_i, \alpha r)}{\hat{n}(p_i, r, \alpha)}$$

In other words, we take the size of our counting neighborhood and compare it to other counting neighborhoods in our sampling area. In our example, p_1 has an MDEF of 2/3, p_2 an MDEF of 0, and p_3 an MDEF of -2/3.

The actual outlier detection process takes the standard deviation of counting neighborhood sizes over the sampling neighborhood, as we see in Listing 11-2.

Listing 11-2. Calculating the standard deviation of MDEF

$$\sigma_{MDEF}\left(p_i,r,\alpha\right)=\frac{\sigma_{\hat{n}}\left(p_i,r,\alpha\right)}{\hat{n}\left(p_i,r,\alpha\right)}$$

Once we have the standard deviation of MDEF, we get to the other input parameter aside from α: k, which represents the number of standard deviations from MDEF before we consider a data point an outlier. To this, we can apply our rule of thumb from Part I of the book: three standard deviations from the mean indicate a data point as an outlier. Papadimitriou et al. describe the formal logic behind this calculation and indicate that when k=3, we can expect fewer than 1% of data points to trigger that threshold if the set of neighborhood sizes follows a normal distribution (Papadimitriou et al., 5).

Now that we have some understanding of how the LOCI algorithm works, we can incorporate it into our multivariate outlier detection model.

Multivariate Algorithm Ensembles

As we begin to integrate LOCI with COF, it is a good idea to take a step back and understand the options available to us with respect to ensembling. The first part of this section will cover the key types of ensembles available to us. The second part of this section will give us a chance to tighten our use of COF via ensemble. The third part of this section will add LOCI to the mix as well.

Ensemble Types

There are two main classes of ensemble: *independent ensembles* and *sequential ensembles*. In Part II, we implemented a form of independent ensemble using weighted shares. The main idea behind an independent ensemble is that we run each test separately, generate each output separately, and combine results at the end. Typically, we return a data point as an anomaly if all—or at least enough—of the tests agree. We can weigh those tests differently based on the quality of different tests. We may, for example, reduce the weight of noisy tests relative to their less error-prone counterparts.

Sequential ensembles, by contrast, stack models like a sieve. The idea here is that the output of one model serves as the input for the next model. In practice, we tend to have the most sensitive (and often noisiest) model go first, giving us a provisional set of outliers. Then, for each additional algorithm, we need only to perform the next set of calculations on the remaining set of outliers—if prior, noisier models think a data point is an inlier, we can accept it as such and save the computational effort. The main benefit to a sequential ensemble is that it can quickly narrow down very large datasets, making it a great approach for enormous datasets. If you are dealing with tens or hundreds of millions of records, a noisy approach should still filter out 90–95% of the data points out of the gate, leaving us with a much smaller fraction for the less noisy but more computationally expensive methods.

The major downside to sequential ensembles is that they tend to require more programming effort, as out-of-the-box implementations of algorithms in libraries like PyOD assume you will care about the entire dataset rather than a small selection of data points. In an effort to minimize complexity, we will again use an independent ensemble for multivariate analysis. This is an area, however, in which it can make sense to invest in developing a sequential ensemble, as the implementations of COF and (especially) LOCI we will use can be time-intensive even for small datasets.

Before we bring LOCI into the mix, it is worth taking a return trip to COF.

COF Combinations

By the end of Chapter 10, we created a multivariate outlier detection system using COF as the centerpiece. COF is a great algorithm, but it does require that callers choose the size of the neighborhood and the cutoff point for what constitutes an outlier. Our implementation partially handled the second item by setting a minimum boundary of 1.35 but left the choice of number of neighbors entirely to the caller. This is a reasonable solution, but we can do better: we can try COF with different combinations of neighbor counts (within reason) and combine the results of each test together as an ensemble. To do so, we need the `combination` module in PyOD.

Listing 11-3 shows our updated run_tests() function. This function now takes the max fraction of anomalies and sends it to COF. We also perform some of the preparatory work, such as converting the encoded values to an array in this function rather than inside check_cof() because we will need the same data for LOCI.

Listing 11-3. The run_tests() function for an ensemble of COF

```python
from pyod.models.combination import aom, moa, average, median,
maximization, majority_vote

# Code elided

def run_tests(df, max_fraction_anomalies, n_neighbors):
  num_records = df['key'].shape[0]
  tests_run = {
    "cof": 1
  }
  diagnostics = {
    "Number of records": num_records
  }
  col_array = df.drop(["key", "vals"], axis=1).to_numpy()

  n_neighbor_range = range(n_neighbors, min(num_records - 5, n_neighbors +
  100), 5)
  n_neighbor_range_len = len(n_neighbor_range)

  labels_cof = np.zeros([num_records, n_neighbor_range_len])
  scores_cof = np.zeros([num_records, n_neighbor_range_len])
  for idx,n in enumerate(n_neighbor_range):
    (labels_cof[:, idx], scores_cof[:, idx], diag_idx) = check_cof(col_
    array, max_fraction_anomalies=max_fraction_anomalies, n_neighbors=n)
    k = "Neighbors_" + str(n)
    diagnostics[k] = diag_idx

  df["is_raw_anomaly_cof"] = majority_vote(labels_cof)
  anomaly_score = median(scores_cof)

  df["anomaly_score"] = anomaly_score
  return (df, tests_run, diagnostics)
```

In this function, we use the `range()` function to generate a series of nearest neighbor counts to try, starting with the user-defined total (or its default of 10) and moving up by increments of 5 to the lesser of 100 plus the original nearest neighbor count or five less than the number of records in the dataset. For example, with 1000 data points and a starting nearest neighbor count of 10, we will get back a total of 21 values, ranging from 10 to 105. By contrast, if we have 20 data points and a starting nearest neighbor count of 4, we will get back three values: 4, 9, and 14. We then build out two-dimensional arrays named `labels_cof` and `scores_cof` to store the results of each test. For each value in the nearest neighbor range, we call `check_cof()`, sending in the column array, max fraction of anomalies, and number of neighbors, storing the results in our array.

After we have taken care of running COF the prescribed number of times, we need to combine the results somehow. PyOD includes several mechanisms for determining the final value for an outlier score or label. Three options are intuitive: `median()`, `average()` (which performs the mean), and `maximization()`. Another useful option is `majority_ vote()`, which we use for determining our label values. Two other interesting options are Average of Maximum (AOM) and Maximum of Average (MOA). AOM and MOA require a number of buckets, which defaults to 5. Then, we randomly allocate results to each of the buckets. For AOM, we find the maximum value in each bucket and calculate the mean across each of those bucket maxima. For MOA, we calculate the mean of each bucket and return the maximum of those values. With both measures, we can smooth out scores across runs for a given data point. For example, if a COF score for a given data point is normally in the 1.1–1.3 range but a single run results in a COF of 7, this can affect our `average()` and `maximization()` functions but will have a very small effect on `median()`, MOA, and AOM. There can be some benefit to using one of AOM or MOA to calculate our response, but median will work well enough and also ensures that we do not need to consider how many buckets we would need for any given dataset. For that reason, we will use the median to calculate our anomaly score.

Moving on, Listing 11-4 shows the new version of `check_cof()`. Compared to the version in Chapter 10, this function exchanges receiving a DataFrame for a column array, includes the max fraction of anomalies as its contamination percentage, and returns arrays rather than columns.

Listing 11-4. A simple function to run our COF test

```
def check_cof(col_array, max_fraction_anomalies, n_neighbors):
  clf = COF(n_neighbors=n_neighbors, contamination=max_fraction_anomalies)
  clf.fit(col_array)
  diagnostics = {
    "COF Contamination": clf.contamination,
    "COF Threshold": clf.threshold_
  }
  return (clf.labels_, clf.decision_scores_, diagnostics)
```

By incorporating a variety of tests covering different numbers of neighbors, we have a more robust calculation of COF, and we can also remove an unnecessary decision from our users in how many neighbors to choose. We are now ready to incorporate LOCI into the mix.

Incorporating LOCI

The changes we made in the prior section to the run_tests() function will allow us to incorporate LOCI in a straightforward manner. Listing 11-5 includes an updated run_tests() function as well as the LOCI process itself.

Listing 11-5. Incorporating LOCI in the multivariate outlier detection process

```
def run_tests(df, max_fraction_anomalies, n_neighbors):
  num_records = df['key'].shape[0]
  if (num_records > 1000):
    run_loci = 0
  else:
    run_loci = 1

  tests_run = {
    "cof": 1,
    "loci": run_loci
  }
  diagnostics = {
    "Number of records": num_records
  }
```

```
col_array = df.drop(["key", "vals"], axis=1).to_numpy()

# COF-specific code removed for clarity
anomaly_score = median(scores_cof)

if (run_loci == 1):
  (labels_loci, scores_loci, diag_loci) = check_loci(col_array)
  df["is_raw_anomaly_loci"] = labels_loci
  anomaly_score = anomaly_score + scores_loci
  diagnostics["LOCI"] = diag_loci

df["anomaly_score"] = anomaly_score
return (df, tests_run, diagnostics)

def check_loci(col_array):
  clf = LOCI()
  clf.fit(col_array)
  diagnostics = {
    "LOCI Threshold": clf.threshold_
  }
  return (clf.labels_, clf.decision_scores_, diagnostics)
```

The first change we see determines whether to run LOCI. As of the time of writing, the implementation of LOCI in PyOD is very inefficient and can struggle with larger datasets. For this reason, if our dataset is too large, we will not run LOCI and will depend on the more efficient implementation of COF. Assuming we do run LOCI, we call the check_loci() function. Note that we do not pass in values of α or k; we leave those at their defaults of 0.5 and 3, respectively. The LOCI algorithm also determines the value of r for us automatically, so we do not need to pass anything in for that parameter. What we get back from the check function includes an array of scores, one outlier score per data point. The result of the median(scores_cof) function call is also an array of scores, containing the median data score per data point across all of our COF tests. We simply add these two arrays together, giving us a combined score with no weighting factor. The intuition here is that our minimum threshold for COF is 1.35 and our threshold for LOCI is 3. We expect both values to be fairly close to one another for inliers and near-outliers, so by adding them together, we let a very strong decision by one model out-vote a tepid response from the other. For example, suppose LOCI returns a score of 2.8, which is close but not quite an outlier. If COF returns a score of 20, well above the 1.35 minimum threshold, the strength of COF's conviction outweighs LOCI's lukewarm judgement.

We can see this play out in Listing 11-6, the updated function to determine outliers. This function previously took three inputs: a DataFrame, our sensitivity score, and the maximum fraction of anomalies we would allow. Now we add two more parameters to the mix: the set of tests run and the sensitivity factors for each test, which are 1.35 for COF and 3 for LOCI.

Listing 11-6. The latest form of our function to determine which data points are outliers

```
def determine_outliers(
  df,
  tests_run,
  sensitivity_factors,
  sensitivity_score,
  max_fraction_anomalies
):
  tested_sensitivity_factors = {sf: sensitivity_factors.get(sf, 0) * tests_
  run.get(sf, 0) for sf in set(sensitivity_factors).union(tests_run)}
  sensitivity_threshold = sum([tested_sensitivity_factors[w] for w in
  tested_sensitivity_factors])
  diagnostics = { "Sensitivity threshold": sensitivity_threshold }

  second_largest = df['anomaly_score'].nlargest(2).iloc[1]
  sensitivity_score = (100 - sensitivity_score) * second_largest / 100.0
  diagnostics["Raw sensitivity score"] = sensitivity_score

  max_fraction_anomaly_score = np.quantile(df['anomaly_score'], 1.0 - max_
  fraction_anomalies)
  diagnostics["Max fraction anomaly score"] = max_fraction_anomaly_score
  if max_fraction_anomaly_score > sensitivity_score and max_fraction_
  anomalies < 1.0:
```

```
    sensitivity_score = max_fraction_anomaly_score
    diagnostics["Sensitivity score"] = sensitivity_score

return (df.assign(is_anomaly=df['anomaly_score'] > np.max([sensitivity_
score, sensitivity_threshold])), diagnostics)
```

We begin the function by calculating the sensitivity threshold, which will be 1.35 if we only ran COF and 4.35 if we ran both COF and LOCI. After that, we calculate the sensitivity score based on the second-largest data point, just as we did in Chapter 10. Then, we determine whether we need to use the sensitivity score or max fraction of anomalies to act as our cutoff, just as we did in Chapter 10. Finally, the return statement creates a new `anomaly_score` column based on whether a data point clears the larger of our sensitivity score and our sensitivity threshold. One final difference in this function is that we return diagnostic data relating to score calculations, allowing a user to diagnose issues.

Test and Website Updates

Whenever we introduce a new algorithm into an ensemble or change the way we calculate things, it makes sense to review the tests and ensure that behaviors change in a way we expect. In this case, we have some tests that expect to return zero outliers, some that return one, and a broader sample dataset with no fixed number of outliers. In this section, we will review those test results and see what fails. Then, we will move to website updates. Note that there is no section on integration tests because those will continue to pass as expected.

Unit Test Updates

There are two changes of note with respect to unit tests. The first is that by running multiple iterations of COF and incorporating LOCI, the number of outliers becomes rather stable: shifts in the sensitivity score and max fraction of anomalies have very little impact until you get to the extremes. On the whole, this is an interesting outcome for us as it means end users should not need to worry about picking the "right" sensitivity score or max fraction of anomalies. The key downside, however, is the loss of control for more advanced users. In the sample input provided with the multivariate test cases, we see five outliers with a sensitivity score of 5. Trying again with a sensitivity score of 25, the number of outliers jumps to eight, and it stays at eight the rest of the way to a sensitivity score of 100. If users anticipate smooth (or at least regular) increases in the number of outliers based on the sensitivity score, they might be confused by this result.

The other change of note comes from the fact that LOCI is more sensitive than COF, at least on our test datasets. For example, the sample inputs with no outliers and one outlier both contain a record whose first input is 87.9 and whose third input is 6180.90844, both of which represent "extreme" values. 87.9 happens to be the largest value we see in the first input and 6180.90844 the smallest value we see in the third, but both are well within 1% of their respective medians. LOCI is capable of spotting this minute difference and flags the data point as an unexpected outlier. To ensure that the tests return the appropriate number of outliers, changing the third value to its median value of 6185.90844 was sufficient to resolve the issue. This level of sensitivity is something to watch out for when using algorithms like COF and LOCI, especially given the first insight around the relative lack of user-controllable sensitivity adjustments.

Website Updates

The website changes in this section are fairly minor and relate to additional debug information we now collect. Listing 11-7 shows the relevant code block.

Listing 11-7. Writing out details if debug mode is enabled

```
if debug:
  col11, col12 = st.columns(2)

  with col11:
    st.header("Tests Run")
    st.write(res['debug_details']['Tests run'])
    st.write(res['debug_details']['Test diagnostics'])

  with col12:
    st.header("Outlier Determinants")
    st.write(res['debug_details']['Outlier determination'])

  st.header("Full Debug Details")
  st.json(res['debug_details'])
```

The first column of our debug menu shows the set of tests run as well as diagnostic information. This diagnostic information includes the number of records in the dataset; the number of neighbors, contamination score, and threshold value for each run of COF; and the LOCI threshold. The second column shows outlier determinants, including our sensitivity threshold and the work we do to calculate sensitivity score. Figure 11-3 shows an example of this using the pre-created sample dataset. Below these two columns, we can see the full debug details as well.

Tests Run

```
▼ {
    "cof" : 1
    "loci" : 1
  }
▼ {
    "Number of records" : 21
  ▼ "Neighbors_10" : {
      "COF Contamination" : 0.11
      "COF Threshold" : 5.283404702391509
    }
  ▼ "Neighbors_15" : {
      "COF Contamination" : 0.11
      "COF Threshold" : 4.142484995035714
    }
  ▼ "LOCI" : {
      "LOCI Threshold" : 3
    }
  }
```

Outlier Determinants

```
▼ {
    "Sensitivity threshold" : 4.35
    "Raw sensitivity score" : 11.518777755984658
    "Max fraction anomaly score" : 7.379453848226641
    "Sensitivity score" : 11.518777755984658
  }
```

Figure 11-3. *The tests we have run and determinants for our outlier cutoff point*

Conclusion

This chapter focused on extending the multivariate outlier detection model in two ways: first, by taking advantage of the `combination` module in PyOD to run the Connectivity-Based Outlier Factor with multiple nearest neighbor counts, and second, by incorporating another algorithm called Local Correlation Integral. The next chapter will extend this model further by incorporating an algorithm entirely different from COF and LOCI.

## CHAPTER 12

Copula-Based Outlier Detection (COPOD)

So far, we have looked at clustering-based models for multivariate outlier detection. In this chapter, we will review a novel nonclustering technique that uses a concept called *copulas*. First, we will define what a copula is and how we can perform outlier detection using it. Then, we will implement application changes to integrate the new technique. After that, we will update our tests and website. Finally, we will wrap up Part III with some concluding notes.

Copula-Based Outlier Detection

In September 2020, Zheng Li and four coauthors released a paper entitled *COPOD: Copula-Based Outlier Detection*. This paper introduced a novel technique for multivariate outlier detection. Prior to this paper, several popular multivariate outlier detection techniques (including LOF, COF, and LOCI) focused on the idea of clustering: calculating the distances between points and calling out those points that are sufficiently distant from their neighbors. By contrast, COPOD relies on a concept known as a *copula*.

What's a Copula?

The term "copula" comes from Latin and means a link or a bond. In language, we use the term "copula" to represent the link between the subject and the predicate, often using a verb like "to be" in English. In statistics, a copula is "a probability model that represents a multivariate uniform distribution, which examines the association or dependence between many variables" (Kenton, 2021).

Breaking that definition into pieces, we start with a dataset containing multiple variables. We may (or may not!) know the distribution of each variable—for example,

K. Feasel, *Finding Ghosts in Your Data*, https://doi.org/10.1007/978-1-4842-8870-2_12

one variable might follow a normal distribution, another a beta distribution, and a third may be uniformly distributed. So far, we have spoken of these variables as if they were entirely independent, but this is often not a good assumption when working with multivariate data. In practice, there tends to be some joint probability distribution, as inputs often have some influence upon one another. For example, the height, weight, age, blood pressure, and shoe size of a person will be related variables, so treating them as entirely independent will leave out important information. The problem that we quickly hit is, how do we model the joint probability distribution of any pair or combination of these variables? This is where copulas come in.

With a copula, we can reduce a joint probability distribution into its *marginal distributions* (also known as *marginals*), which are independent and therefore not correlated between variables. The copula is a function that allows us to make this transformation from a single joint distribution to a coupling of marginal distributions.

Note   Copula theory is very useful in certain avenues of statistics, but a detailed understanding of the topic is beyond the scope of this book. For an intuitive, visual-heavy approach to understanding copula functions, see Wiecki (2018). For a more detailed survey of copula theory, see Durante and Sempi (2010) as well as Pakdaman (2011). For our purposes, this high-level understanding should suffice.

Intuition Behind COPOD

Li et al. released a paper in September of 2020 introducing a new technique, *Copula-Based Outlier Detection* (COPOD). COPOD takes advantage of copulas to break down the relationships between multiple variables, even in cases in which we do not know the underlying distribution of the individual variables themselves. There are three stages for detecting outliers using COPOD (Li et al., 2). The first stage is to compute the *empirical cumulative distribution function* (also known as an empirical CDF or ECDF), which describes the probability at each point that a random variable will be less than or equal to that point. In Chapter 3, we introduced the concept of a *probability distribution function* (PDF), including an image like that in Figure 12-1. To oversimplify the explanation a bit, the height of the curve represents the probability some random variable drawn from this distribution takes on the value on the X axis.

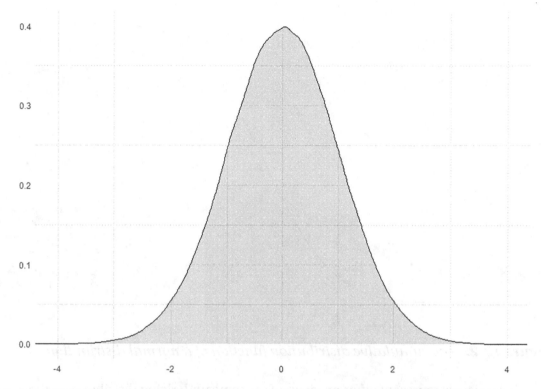

Figure 12-1. *The probability distribution function of a normal distribution*

The probability distribution function is a "point" measure. The cumulative distribution function of the same graph, which appears in Figure 12-2, represents a "flow" measure. In other words, what is the probability that we will draw a point from this distribution that is less than or equal to a given value on the X axis?

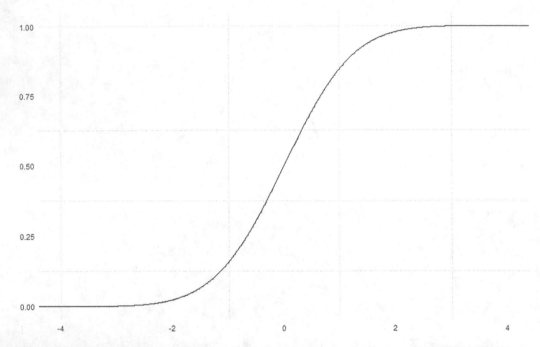

Figure 12-2. *The cumulative distribution function of a normal distribution*

These visuals represent different views of the same underlying distribution—the cumulative distribution function is the integral to the probability distribution function over the range negative infinity to x (or, in other words, the probability distribution function is the first derivative of the cumulative distribution function, assuming the CDF is differentiable).

Once we have the empirical CDF for each variable, the second stage of the process is to determine an empirical copula function that translates the joint probability distributions of variables into marginal distributions.

The final stage of the process uses the empirical copula to approximate the *tail probability*, which is the probability that a sampled data point would be greater than (for a right tail) or less than (for a left tail) some specified value. We do this for each variable and multiply the probabilities together to create a multivariate tail probability—something we can do because the copula provides us marginal distributions, which are by definition independent from one another.

Given a tail probability, Li et al. show how to use this to estimate the likelihood of a data point being an outlier. The authors have also provided an implementation of their algorithm in Python, which they have incorporated into the PyOD library.

Implementing COPOD

Similar to LOCI in Chapter 11, COPOD does not have any required user-defined parameters. We can see this in Listing 12-1 as we implement a COPOD check.

Listing 12-1. The function to run COPOD testing against our multivariate dataset

```
def check_copod(col_array):
  clf = COPOD()
  clf.fit(col_array)
  diagnostics = {
    "COPOD Threshold": clf.threshold_
  }
  return (clf.labels_, clf.decision_scores_, diagnostics)
```

COPOD's implementation is quite fast, so we are safe to run this no matter the number of records. Listing 12-2 shows a simplified version of the run_tests() function that now includes COF, LOCI, and COPOD as algorithms in our ensemble.

Listing 12-2. The run_tests() function now includes COPOD.

```
def run_tests(df, max_fraction_anomalies, n_neighbors):
  num_records = df['key'].shape[0]
  if (num_records > 1000):
    run_loci = 0
  else:
    run_loci = 1

  tests_run = {
    "cof": 1,
    "loci": run_loci,
    "copod": 1
  }
  diagnostics = {
    "Number of records": num_records
  }

  col_array - df.drop(["key", "vals"], axis=1).to_numpy()
```

```
# COF and LOCI code elided

(labels_copod, scores_copod, diag_copod) = check_copod(col_array)
df["is_raw_anomaly_copod"] = labels_copod
diagnostics["COPOD"] = diag_copod
df["anomaly_score_copod"] = scores_copod
anomaly_score = anomaly_score + scores_copod

df["anomaly_score"] = anomaly_score
return (df, tests_run, diagnostics)
```

Now that we have a third algorithm in our ensemble, we will need to adjust weights to determine our cutoffs. Li et al. suggest using a threshold such as `-ln(0.01)` = `4.61` (Li et al., 4) per dimension. In our case, we will use a more aggressive score of 2.3 (which is `-ln(0.10)`) above the median as our cutoff for COPOD. We can see the final form of `determine_outliers()`, which incorporates COPOD, in Listing 12-3.

Listing 12-3. The `determine_outliers()` function now includes calculations for COPOD.

```
def determine_outliers(
  df,
  tests_run,
  sensitivity_factors,
  sensitivity_score,
  max_fraction_anomalies
):
  tested_sensitivity_factors = {sf: sensitivity_factors.get(sf, 0) * tests_
  run.get(sf, 0) for sf in set(sensitivity_factors).union(tests_run)}
  median_copod = df["anomaly_score_copod"].median()
  sensitivity_threshold = sum([tested_sensitivity_factors[w] for w in
  tested_sensitivity_factors]) + median_copod
  diagnostics = { "Sensitivity threshold": sensitivity_threshold, "COPOD
  Median": median_copod }

  second_largest = df['anomaly_score'].nlargest(2).iloc[1]
  sensitivity_score = (100 - sensitivity_score) * second_largest / 100.0
  diagnostics["Raw sensitivity score"] = sensitivity_score
```

```
max_fraction_anomaly_score = np.quantile(df['anomaly_score'], 1.0 - max_
fraction_anomalies)
diagnostics["Max fraction anomaly score"] = max_fraction_anomaly_score
if max_fraction_anomaly_score > sensitivity_score and max_fraction_
anomalies < 1.0:
  sensitivity_score = max_fraction_anomaly_score
  diagnostics["Sensitivity score"] = sensitivity_score

return (df.assign(is_anomaly=df['anomaly_score'] > np.max([sensitivity_
score, sensitivity_threshold])), diagnostics)
```

Our calculation for `tested_sensitivity_factors` remains the same as in Chapter 11, iterating over each test and multiplying the weight by the value (0 or 1) of `tests_run`.

The next step calculates the median for COPOD. Unlike LOCI and COF, which use absolute measures, COPOD's score is a relative score from the median, meaning we need to perform this calculation and add it to the sensitivity threshold. The net result is that our minimum sensitivity threshold is 3.65, adding the 1.35 sensitivity factor for COF with the minimum 2.3 sensitivity factor for COPOD. If we include LOCI as well (as we do when we have no more than 1000 observations), our minimum sensitivity threshold will be 6.65. After that, the rest of the function remains unchanged.

Now that we have covered all of the necessary changes to the code, let's continue on to tests and website updates.

Test and Website Updates

As we have seen several times throughout the book, introducing a new algorithm into an ensemble will affect existing tests. In this section, we will review the net effects on unit and integration tests. Because this is the final chapter in which we work on multivariate outlier detection, we will also extend the website to make it a bit more user-friendly.

Unit Test Updates

When running unit tests, the big change comes with the number of outliers in the sample input. Interestingly, the range of valid values has shrunk considerably, as we can see in Table 12-1.

Table 12-1. *Number of outliers for differing sensitivity scores given a max fraction of anomalies of 1.0*

Sensitivity Score	Ch. 10	Ch. 11	Ch. 12
100	11	8	6
50	11	8	6
40	11	8	6
25	4	8	6
5	2	2	5
1	1	2	3

In Chapter 10, we saw a range from 1 to 11 outliers depending on the sensitivity score, with a "dead zone" from 40 to 100 in which the number of outliers did not change. In Chapter 11, we saw either 2 or 8 outliers, with the cutoff happening somewhere between 25 and 5. In this chapter, the minimum number of outliers has increased to 3, and the maximum has dropped to 6. This has further reduced the relative importance of sensitivity score.

Integration Test Updates

When running the integration tests for Chapter 12, we end up with a test failure that is actually a positive outcome. In Chapter 10, we introduced an integration test that includes three outliers, one of which is fairly extreme and the other two reasonably small. In the Postman tests, this integration test is called `Ch10 - Multivariate - Debug - Three Outliers, 2S1L - Two Recorded`. We noted near the end of Chapter 10 that only the two largest outliers were caught by COF using our input parameters. In Chapter 11, the ensemble of COF and LOCI did not change the results of this integration test. With the introduction of COPOD, the anomaly score for the third data point increases enough to clear the threshold for detection. Figure 12-3 shows this result in visual form.

Anomaly score per data point

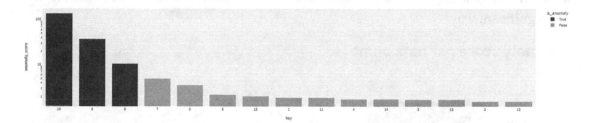

Figure 12-3. *Three outliers are detected, whereas previously, COF would detect only two*

This is a positive outcome for us, as we now are able to show that one large outlier and one small-to-medium outlier will not necessarily overwhelm a small outlier.

Website Updates

Before we close the books on multivariate outlier detection, there is a small amount of functionality we should add to the Streamlit site. Listing 12-4 shows the first set of changes, in which we include the separate anomaly score components in the hover data for each point as well as our results table. This will allow us to see which algorithms have the most to say about a given data point.

Listing 12-4. Include separate anomaly score components for COF, LOCI, and COPOD

```
g = px.bar(df, x=df["key"], y=df["anomaly_score"], color=df["is_anomaly"],
color_discrete_map=colors,
        hover_data=["vals", "anomaly_score_cof", "anomaly_score_loci",
        "anomaly_score_copod"], log_y=True)
st.plotly_chart(g, use_container_width=True)

tbl = df[['kcy', 'vals', 'anomaly_score', 'is_anomaly', 'anomaly_score_
cof', 'anomaly_score_loci', 'anomaly_score_copod']]
st.write(tbl)
```

We can see the result of this change in Figure 12-4. For the most part, COPOD seems to drive the overall anomaly scores, though the result is not quite as drastic as it may first appear. We do need to factor in the COPOD median of 6.08 and the COPOD sensitivity factor of 2.3, meaning that the COPOD score must be above 8.38 to move the needle in any significant way.

Anomaly score per data point

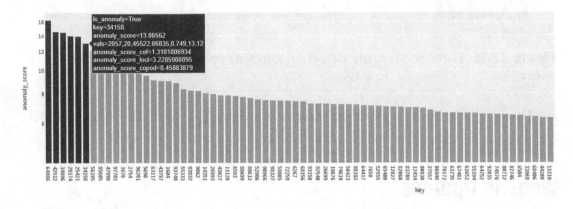

	key	vals	anomaly_score	is_anomaly	anomaly_score_cof	anomaly_score_loci	anomaly_score_copod
40	64886	[6061,5,4680.25499,0.552,3.32]	16.2796	true	0.9216	7.0000	8.3581
79	42512	[2640,20,49648.53389,0.865,18.17]	14.5457	true	1.3456	3.0099	10.1902
17	34886	[2110,20,47921.24632,0.812,17]	14.4443	true	1.3157	3.7360	9.3927
24	29174	[1726,20,47167.66052,0.78,13.62]	13.9702	true	1.3170	3.5560	9.0971
90	25421	[1468,20,46906.47825,0.789,14.7]	13.9104	true	1.3177	3.5015	9.0912
10	34158	[2057,20,45522.06835,0.749,13.12]	13.0056	true	1.3182	3.2286	8.4588
69	56195	[4833,20,42826.20755,0.726,17.2]	12.1313	false	1.3693	3.0022	7.7598
55	85605	[7248,16,4803.94926,0.608,4.86]	11.5109	false	0.9616	3.9725	6.5768
19	47898	[3300,4,5699.15996,0.273,37.04]	11.2979	false	0.8049	0.6917	9.8013
52	97702	[8658.7,17132.85035,0.99,51.12]	11.1261	false	1.1313	-0.4404	10.4352

Figure 12-4. *Using a large, artificially generated dataset (`sample_input` in the multivariate unit test suite), we can see that each data point has a breakdown of its anomaly score components, covering COF, LOCI, and COPOD separately*

The other change involves accepting a list of data. For univariate data, we accept a list of data points in Python format, something like [1, 2, 3, 4, 5]. We then convert this to key-value pairs and send the results to the outlier detection engine. For multivariate data, it would be nice to accept data in the same format as our unit tests, in which each row is a list that contains two elements: a key and a list of values. An example of one data point in this format is ["key1", [1, 2, 3, 4, 5]]. We could then create the dataset by generating a list of these data points. Once we do that, we should be able to parse the list, translate it to JSON, and send that result to the outlier detection engine. Listing 12-5 shows how to do just that using Python's Abstract Syntax Tree (ast) module.

Listing 12-5. Create an abstract syntax tree to parse a list of lists and convert results to appropriate JSON.

```
@st.cache
def convert_multivariate_list_to_json(multivariate_str):
  mv_ast = ast.literal_eval(multivariate_str)
  return json.dumps([{"key": k, "vals": v} for idx,[k,v] in
  enumerate(mv_ast)])
```

Using the literal_eval() function, we evaluate the string and build out a list of lists. The json.dumps() function translates this list into valid JSON in the format we require by enumerating over each data point in the dataset, parsing out the key and value, and tagging them with the appropriate names.

To operate this function, we need a basic check that runs input_data = convert_multivariate_list_to_json(input_data) whenever we are using the multivariate method and the user has selected the checkbox to convert list data to JSON. Figure 12-5 shows an example of this, taking the sample_input unit test dataset as the input and building a proper JSON body for our outlier detection engine.

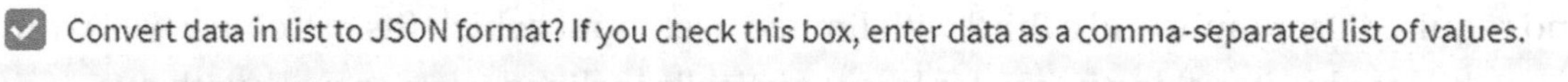

☑ Convert data in list to JSON format? If you check this box, enter data as a comma-separated list of values.

Data to process (in JSON format):

```
[["1604", [87,16,6184.90844,0.771,11.72]],
 ["91849", [7921,12,6337.69829,0.919,11.55]],
 ["55194", [4497,5,5639.15773,0.678,4.71]],
 ["63735", [5969,18,29296.31524,0.605,26.42]],
 ["66847", [6166,18,29250.95360,0.616,25.23]],
 ["40687", [2508,12,6002.59147,0.847,9.34]],
 ["68406", [6276,18,30314.86144,0.639,23.94]],
 ["58621", [5637,13,28236.24848,0.760,3.70]],
 ["26993", [1580,7,14753.01818,0.844,46.33]],
 ["72869", [6529,8,12119.75085,0.599,20.10]],
 ["34158", [2057,20,45522.06835,0.749,13.12]],
```

Detect!

Figure 12-5. *Convert a list of data point lists into valid input for outlier detection*

Conclusion

Over the course of this chapter, we learned just enough about copulas to understand the basic workings of the Copula-Based Outlier Detection (COPOD) method. We then incorporated COPOD into our multivariate outlier detection ensemble to good effect. We concluded by making some small but useful improvements to the accompanying outlier detection application.

Before we move on to Part IV and time series analysis, we should take a moment to reflect on what we've been able to accomplish and how we could make the current multivariate outlier detection system better. Throughout Part III, we have built an independent ensemble of three algorithms: Connectivity-Based Outlier Factor (COF), Local Correlation Integral (LOCI), and COPOD. These three algorithms work in different ways to discover if some data point appears to be an outlier, with two of the algorithms using density-based approaches and the third a distributional approach using copula functions. We have seen that the combination of these three algorithms provides us a rather stable base for determining outliers, meaning we do not see radical differences in the number of data points marked as outliers given different sensitivity scores or numbers of nearest neighbors. We also have the benefit of choosing two algorithms with no required user input and a third with limited user input, which fits extremely well with our philosophy of making it easy for less statistically inclined users to work with our service.

There are dozens of other algorithms available for multivariate outlier detection we could investigate, though each additional algorithm adds new complexities and should be carefully evaluated before addition. COF and LOCI (specifically ALOCI, a linear approximation of the LOCI algorithm not available in PyOD) work fairly well based on Mehrotra's research (Mehrotra et al., 116–117), which came out before the COPOD paper. With that as our starting point, more is not necessarily better. To ensure that we make solid decisions on further algorithmic choices, a deeper analysis over additional, labeled datasets would be critical.

This deeper analysis would also help us fine-tune the weights. The LOCI algorithm comes with strong hyperparameter guidance from its authors, but COF and COPOD do not. We made reasonable decisions for each of these weights, but additional datasets could allow us to tweak these measures for better results.

We are now done with multivariate outlier detection and will shift to Part IV, time series analysis. In the next chapter, we will get an overview of time series problems.

PART IV

Time Series Anomaly Detection

Up to this point, time has not been a factor in our anomaly detection engine. In Part IV, we introduce a new wrinkle to the problem by incorporating time into the process. Chapter 13 shows us why time is such a special consideration in outlier detection and shows us several techniques for solving time series problems. From there, Chapter 14 provides us information on a concept known as change point detection, one efficient method for time series outlier detection. In Chapter 15, we make the transition from single-series time series problems to multi-series time series problems. We get an idea of what multi-series time series is, how it differs from single-series time series, and what assumptions we need to make in order to solve multi-series problems. After that, we implement one technique for time series analysis called Standard Deviation of Differences (DIFFSTD) in Chapter 16 and see how we can use this pairwise comparison technique to solve multi-series problems. Finally, Chapter 17 covers another time series technique called Symbolic Aggregate Approximation (SAX).

## CHAPTER 13

Time and Anomalies

Now that we have entered Part IV of this book, it is time to introduce an important topic in anomaly detection: time series anomaly detection. Just as with Parts II and III, we will break up time series analysis based on the number of features: first, we will deal with single-series time series anomaly detection; then, after we have a model in place for single-series time series anomaly detection, we will deal with the multi-series case.

This chapter will start by providing the basis behind single-series time series analysis. We will gain an understanding of why time series is such an important topic. We will also learn how time series analysis changes our thinking on anomaly detection. After that, we will look at some of the most common time series analysis problems and will wrap up with a discussion of what actually constitutes an outlier when performing time series anomaly detection.

What Is Time Series?

At its core, *time series data* is data with some combination of date and time in addition to at least one feature. Single-series time series analysis means that all of the data points we collect represent findings from the same phenomenon at different points in time. We will also refer to any combination of date and time as "time" to keep things simple—a review of stock market data by date alone will still constitute time series analysis. Figure 13-1 shows an example of time series data, charting the Earned Run Average of Greg Maddux over his career in Major League Baseball. In this case, the *granularity* of our time series data is annual, as this is the most common grain for Earned Run Average.

K. Feasel, *Finding Ghosts in Your Data*, https://doi.org/10.1007/978-1-4842-8870-2_13

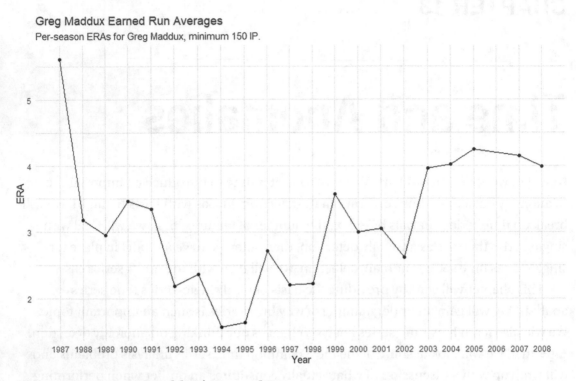

Figure 13-1. *Greg Maddux's Earned Run Averages per year*

Time series analysis is important for a few reasons. First, time is a special variable in that it always moves in one direction—although if I do get access to a time machine that allows me to travel back in time, I'll be sure to update this comment in the book!

The second reason time series analysis is important is that not only it is time linear but it also acts as a contributing cause to processes. In other words, processes take time as one input: if we decide to act today, something may come to fruition at some point in the future. Our language is even full of terms like "fruition," which is the process, over time, of bearing fruit. The fact that most processes take time as one input component makes time an interesting feature by which we can analyze our data.

This leads to the third reason: many business problems have a time component. For example, we may track revenue growth over time with the expectation that it should, on average, grow at some fixed percentage rate. Another example is that we may estimate demand for various products in our catalog so we know how many to manufacture or purchase beforehand.

That said, time series analysis is certainly not the only type of analysis we could perform, and this should not minimize the work we have put into *static* (i.e., time-free) outlier detection techniques so far. We should use time series analysis whenever it makes sense, but incorporating an unnecessary time feature simply to make something into time series analysis ultimately harms our analysis of a phenomenon.

Time Series Changes Our Thinking

The special nature of time as an input feature has several ramifications for how we should analyze datasets with a time series component. In this section, we will look at several ways in which time series analysis differs from classical data analysis.

Autocorrelation

Time series data exhibits *autocorrelation*, otherwise known as *serial correlation*. Autocorrelation means that the value of a data point at time T is directly affected by its point at time T-1 (or, more generally, T-k for some k >= 1). For example, if you cross a large but busy room, your position three seconds in will depend on where you were a few seconds ago. The intuition behind this is that we are incapable of instantaneous travel and instead may only travel some maximum distance over the course of a few seconds. Ignoring the factors that may drive what that maximum distance can be (e.g., height, gait, injuries, and so on), we know that if a person can travel a maximum distance of d over the course of three seconds and if they are at point x at the end of the three seconds, we can draw a circle of radius d around point x and know that our person was within that radius three seconds ago. Figure 13-2 shows this principle in action.

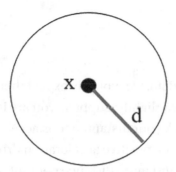

Figure 13-2. *Given the current position of a person and the maximum distance that person could travel in a fixed time frame, we know they must have been within the circle at the beginning of the time frame*

It is worth belaboring this point because the same Figure 13-2 can give us an idea of where a person could be at the end of the next time frame of three seconds: somewhere within a circle of radius length d from point x. If we have sufficient information about where that person was at different points in time, we might come up with a result like Figure 13-3, in which we see a person's motion over time from t-5 until now (time t), as well as the possible range of where they could be at time t+1.

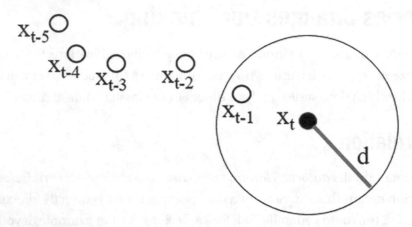

Figure 13-3. *A person's movement over the course of six time periods. This image also includes where the person could go over the next time period*

Admittedly, Figure 13-3 ignores some of the laws of physics around objects in motion: a person moving at some nonzero velocity is heading in some direction. Assuming they can maintain that same velocity, the person may continue moving in that same direction and travel distance d. If the person wants to reverse course, it will take some time to slow down, reverse course, and accelerate in this new direction. This leads to our next point.

Smooth Movement

When dealing with time series data, we typically expect there to be fairly smooth movements. Switching examples slightly, suppose you are in a stopped car on a straight track and we use sensors to track your instantaneous acceleration, velocity, and distance. As you depress the accelerator, the positive acceleration drives increased velocity, and this translates to moving more and more distance per unit of time. Even so, we expect this movement to be somewhat smooth—we don't expect you to start with a speed of 0 MPH, immediately jump to 80 MPH, drop down to 12 MPH, increase to 95 MPH, and so on.

In fact, if our sensors did report this, we would expect sensor failure rather than the idea of sudden speed changes. This is because vehicle acceleration, over the appropriate time frame, is *continuous*: there are no jumps or discontinuities in the data; instead, we can plot a smooth line as long as we have data at the appropriate granularity. That line may certainly move up or down depending on what actions you take, but until we develop some form of instantaneous travel, movement is continuous.

The extent to which we expect to see smooth movement will depend upon the nature of our scenario and the grain of our data. Tracking vehicle acceleration, velocity, and distance per millisecond will lead to rather smooth data, even for cases like rocket boosters taking off into space. By contrast, tracking stock market prices at a daily grain will lead to bigger jumps between data points. This movement leads to the third point.

The Nature of Change

As soon as we start dealing with time series data, we build up an expectation of change over time. That is, if something does not change over some relevant time frame, then we probably do not have time series data. If we review one set of test scores for all 4th graders in a state and look for outliers, this is not time series data: the scores themselves will not change tomorrow or next month. By contrast, if we wish to track growth in students over the course of a year and have tests at periodic intervals, we now have a time series problem because we expect some sort of change over our relevant time interval of one school year.

There are four types of changes that are important to us. The first of these is *trend*, which is a broad movement upward or downward over the course of the time series analysis. If we track the price of some product over time, there are a few factors that will drive the trend. One such factor is inflation, the increase in the money supply relative to goods and services in the market. Another factor is changes in demand over time. Some products become more popular, and that shift in demand can drive increased prices. For example, if a local sports team becomes one of the best in the league, we could expect demand (and price) of tickets to watch them play to increase. This particular example is exacerbated by our third trend: changes in supply. In the sports team example, the number of seats available at the home stadium is pretty much fixed—teams can do some things like block off certain sections of seating to reduce supply, but they will have fewer avenues of increasing seating in the short term. By contrast, as people adopt some new piece of technology, additional competitors come into the market, and established players ramp up production, increasing the level of supply and thereby affecting the price.

The second type of change that is important is *seasonality*. Seasonality deals with periodic swings in our time series value. Sales of beach umbrellas are likely to be much higher in the spring and summer months than in the middle of winter. Similarly, if everybody were paid once on the 15th of each month, we would expect a spike in goods sold on the 16th–18th of each month compared to the 12th–14th, as people are newly flush with cash vs. scrimping to make ends meet. The *periodicity* of this seasonal data tells us how often we run through one seasonal fluctuation. With beach umbrellas, our period is annual. With the contrived wages example, the period would be monthly. If we tracked the amount of energy a solar panel generates per hour in a desert environment, we would expect daily seasonality: the amount of energy generated increases through the course of the day and then decreases into the evening, with this cycle beginning anew the following day.

The third type of change is *cyclical variation*. This is similar to seasonality except that it tends to run over longer time frames and is not as regular. For example, suppose we are trying to predict the amount of rainfall on some tropical island. The seasonal component to this would be the time of year: we might have a rainy season during the winter and a hot season during the summer. Suppose we also know about an approximately 13-year cycle in which rainfall ebbs and wanes. One cycle might last sixteen years and another nine years, so it is not really periodic in the way a computer can easily discern. This can make it very difficult for automated time series analysis to pick up on nonseasonal cycles.

The fourth type of change is *noise*. These are irregular changes that are neither changes in trend nor seasonal behavior nor part of a cycle. With noise, we do not know the cause of the change based solely on the information we have available.

Figure 13-4 shows an example of time series analysis using Consumer Price Index data. The specific series covers electricity and piped gas costs for urban consumers in the Dallas, Fort Worth, and Arlington, Texas area from January of 1997 through July of 2021. Consumer Price Index data is a relative measure of pricing, and this particular example uses the average price from 1982 to 1984 as its baseline of 100. Differences from 100 indicate percentage point differences, so a value of 200 means that prices were twice as expensive as in the baseline.

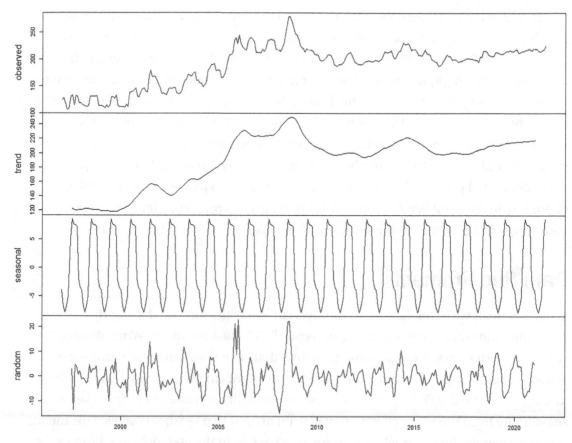

Figure 13-4. *Decomposing one CPI time series into its trend, seasonal, and noise components*

The first chart in the series shows the actual CPI values, which started at 125.20 in July of 1997 and ended at 224.73 in July of 2021. The second chart shows the trend, which steadily increases until its peak of 279.90 in August of 2008. Then, once the housing market collapsed, the trend dropped to the 180–200 range and has slowly increased since. The third chart shows the seasonal effect. Prices tend to be higher in summer than in winter, which makes intuitive sense: winters in the Dallas-Fort Worth-Arlington area tend to be quite mild, but summers get very hot. As a result, people run air conditioning heavily during the summer, leading to a spike. There is a minor spike in winter for heating but lower than the summer height. In spring and fall, when neither heat nor air conditioning is as important, prices decline. Looking closely at the chart, we can see a range of approximately 15 or so CPI points between the lows of spring and the height of summer, meaning that seasonality plays an important role in our analysis but is not the primary driver.

The final chart shows *noise*, the "random" component in the time series analysis. This is all of the change that is left over after trend and seasonality. We expect this to average out at 0 and not follow any perceptible pattern. If we do see a strong pattern in the noise component, it is a good indicator that we failed to account for an important factor. For example, if we have cyclical variation, we might observe it here in the noise component or in the trend component. Another common example of this is if we get the seasonal period wrong. If we assume monthly seasonality—that electricity prices jump up at the beginning of the month but drop at the end—then we will miss the quarterly trend described previously. Time series libraries can help us detect the most likely candidate for seasonality, but the best bet is to have a strong knowledge of your domain, including what drives periodic behavior. We're also going to need more data.

Data Requirements

When we dealt with univariate and multivariate datasets, some of the algorithms we chose had minimum size requirements, typically 15–30 data points. When dealing with time series data, we want more data than that. The intuition behind this desire is that things change over time, so we need more points to understand those changes and determine if the change is due to trend, seasonality, or noise. In order to separate seasonality from the other effects, we will want at least two seasonal cycles. This means that if we expect the seasonal cycle to run one year, as in the electricity scenario in Figure 13-4, we would need to collect at least two years' worth of data. Ideally, we would have four to six seasonal cycles to ensure not only that we capture seasonal effects but also understand if the phenomenon we are observing contains *epicycles*, cycles within cycles. For a contrived example of epicyclic behavior, suppose we chart international tourism from European travelers. We might expect an annual cycle peaking in the summer, as people take lengthy vacations and travel to exotic destinations. Every four years, we see an even larger burst of international tourism due to the summer Olympics. If we model behavior solely on an annual basis, we might miss an upward spike in "noise" every four years as a result. Instead, knowing that we have two seasonal effects, we could model that—but only if we have at least eight years, which would give us two Olympics to observe.

Time Series Modeling

Now that we have an idea of how time series analysis differs from non-time series analysis, let us move on to a brief review of four time series techniques. For each technique, the intent is to make a prediction of future behavior based on prior data movements. Importantly, these are not techniques for finding outliers, but rather they are intended for estimation purposes; still, they help us understand the nature of time series analysis better and will help us get to proper outlier detection techniques in the following chapters.

(Weighted) Moving Average

Moving Average is a simple yet surprisingly effective class of technique for analyzing time series data and making predictions on future data. One common formulation of Moving Average is *Symmetrically Weighted Moving Average* (SWMA), available in Listing 13-1. In this formulation, we take the moving average of a fixed number of periods, weighting the middle periods more heavily than the most recent and most distant periods.

Listing 13-1. Calculating the Symmetrically Weighted Moving Average for a dataset

$$x_t = \frac{1}{6}x_{t-1} + \frac{2}{6}x_{t-2} + \frac{2}{6}x_{t-3} + \frac{1}{6}x_{t-4}$$

SWMA relies on the idea that most time series operations tend to behave cyclically, and so it makes sense to temper the results of the most recent period with that of prior periods. In addition to SWMA, there are also other Moving Average models that put a larger emphasis on more recent periods, include additional prior periods, or otherwise tweak the standard formula.

Exponential Smoothing

Exponential smoothing is a technique that builds up estimations over time, placing a much higher weight on recent data and increasingly less on prior data. Listing 13-2 shows the two-part formula for calculating one variant called *simple exponential smoothing*, starting with a "base case" for the first time period and then a recursive function as the general case.

Listing 13-2. The formula for exponential smoothing

$$s_0 = x_0$$

$$s_t = \alpha x_{t-1} + (1-\alpha)s_{t-1}, t > 0, 0 < \alpha < 1$$

For the first period, we do not have any prior data, and so our estimate s_0 of x at time 1 is the value of x itself at time 0. After we have at least two periods, we can begin to build up a table of calculations over time. Table 13-1 includes the first seven months of CPI data from Figure 13-4, showing each step of the simple exponential smoothing process assuming a smoothing factor of α = 0.3.

Table 13-1. *A simple walkthrough of estimating the next month's CPI data using simple exponential smoothing*

CPI	αx_{t-1}	$(1 - \alpha)s_{t-1}$	s_t
125.20			125.20
128.60	0.3*125.20=37.56	0.7*125.20=87.64	125.20
117.60	0.3*128.60=38.58	0.7*125.20=87.64	126.22
106.60	0.3*117.60=35.28	0.7*126.22=88.35	123.63
107.60	0.3*106.60=31.98	0.7*123.63=86.54	118.52
130.80	0.3*107.60=32.28	0.7*118.52=82.96	115.24
133.40	0.3*130.80=39.24	0.7*115.24=80.67	119.91

In this case, simple exponential smoothing consistently lagged behind CPI in terms of price changes, dropping more slowly during downswings but also increasing more slowly during upswings. This particular variant of exponential smoothing also has no ability to anticipate seasonal behavior, meaning that every summer upswing will catch the simple exponential smoothing algorithm unawares. In fairness, there are other variants of exponential smoothing that do support trend and seasonality, though we do not cover any of them in this book.

Autoregressive Models

The final class of algorithm we will look at incorporates a technique known as *autoregression*. Autoregression starts with the concept of *autocorrelation*, the predictability of a value at some time knowing only its value at previous times. Autocorrelation ranges from +1 to -1, where +1 means the prior data points have a perfect positive correlation with the current data point and -1 means there is a perfect negative relationship. An autocorrelation value of 0 indicates that the prior value has no bearing on the current value. Autoregression (AR) itself is a technique that observes and uses the relationship between the current data point and some number of prior data points. With autocorrelation, we assume that there are only two components: weighted information of some number of past values and a random error term.

In addition to autoregression, we also have the notion of a *moving average* (MA) model. This is a bit different from our discussion of weighted moving average algorithms before. Weighted moving average algorithms like SWMA give us a *smoothed* estimation, averaging results from multiple time periods to smooth out peaks and troughs. By contrast, MA modeling uses the idea of autocorrelation and helps us determine if the "random" error term is actually random, or if there is relevant information hidden within. Another way to put this is that smoothed moving average calculations create a weighted moving average of the actual values at prior times, whereas a moving average model creates a weighted moving average of the error terms.

We can then combine autoregressive and moving average models to form ARMA models. In an ARMA model, we review the impact of previous lags in data (the AR component) as well as previous residuals (the MA component). One requirement we have not covered yet about AR, MA, and ARMA models is that the data must be *stationary*. A time series is stationary if the properties of the series do not depend on the time in which you observe the series. The key factors we would look at here are the mean, variance, and autocorrelation factor (ACF) for a certain slice of data. If we return to the Dallas-Fort Worth-Arlington price of electricity example, we know that the average price is increasing over time. This increase over time causes the data to be *nonstationary*, as our mean in the year 2007 will differ from the mean in the year 2021 and thus when we observe the data matters. If we simply tried to apply an AR, MA, or ARMA model to this dataset, we could end up drawing invalid conclusions as a result. To get around this, we use a concept known as *differencing*. The simplest form of differencing involves subtracting the prior data point's value from the current value. We can perform multiple levels of differencing to estimate quadratic or higher-order trends. Once we have incorporated differencing, we have an AutoRegression model with Moving Average that uses an Integrated (differencing) technique, or ARIMA. ARIMA is a powerful technique for analyzing time series data and generating predictions.

Now that we have an idea of how several time series modeling techniques work, let's now consider what it means for a time series data point to be an outlier.

What Constitutes an Outlier?

So far, most of our time series analysis has focused on predicting the next value (or the next several values) from a single time series. In this section, we will begin categorizing what constitutes an outlier in a time series dataset. We will also broaden the terms of time series analysis to move from one single dataset to multiple time series in a single dataset. An example of this is stock market data: in the same market, we have a variety of stocks whose prices move up and down. For this section, let's model behavior in an office building in which each person uses a key card to enter and we are able to log who enters at what point.

Local Outlier

We can define a *local outlier* as a data point in one time series whose noise component is unexpectedly large. For example, suppose we have an employee who typically scans in three times a day: once at the beginning of the day, once after lunch, and once after a short afternoon break. Occasionally, the person might skip the break or may take two breaks, so we normally see two to four scans on a daily basis, and this has been the case for several years (implying stationarity in the dataset) with no seasonal component. Today, this person has scanned in eight times. This number is unexpectedly high, with an error term 5 higher than the trend of 3. There is necessarily an external explanation for this: it might be that the employee was helping to bring in loads of supplies for an office party, a scanner failure that caused double scanning on each entry, or the employee has copied their key card and given the copied cards to compatriots in an attempt to ransack the office after hours. We will not be able to discern the specific cause of this difference strictly from the time series dataset, but we can tell that there was a difference significant enough to warrant our attention and that constitutes an outlier.

Behavioral Changes over Time

The second type of outlier we might care about is a behavioral change that happens over time. Due to a series of strict deadlines, our employee is no longer able to go out to lunch and must eat at their desk. They also reduce the number of afternoon outdoor breaks almost to zero. Now, instead of scanning in three times a day on average, they scan in one time a day on average and stay in the building the entire day. The trend before this change differs from the trend after this change, and the point in which the change occurs—also known as the *change point* and the topic of the next chapter—is of interest to us.

Local Non-outlier in a Global Change

The third class of outlier is not as straightforward as the first two. Suppose our employee continues to scan in three times a day on average. At the same time, the average number of scans per day for all other employees drops from 2.85 to 0, as the office shuts down for a multiyear period due to unspecified reasons. Our employee's behavior has not changed, but their behavior *relative to the rest of the organization* has changed significantly, moving them from an employee near the average to one well outside the norm.

Differences from Peer Groups

The fourth class of outlier depends on knowing the relationships between individual time series in a set. Going back to our employee example, we could think about formal peer groups, such as people who work in the same department. We might also think of informal peer groups, like a group of people who go to lunch together every day or a group of people who take smoke breaks at similar times. If we know enough about these formal or informal relationships, we could group together these time series in the dataset and understand the extent to which they move in conjunction. Then, if we see a change in behavior from one series in the group, we might consider this outlier behavior even if the series itself does not drastically change. A concrete example of this is that our employee who normally scans in three times a day is part of a common group. The common group, due to whatever reason, drops down to approximately 2.4 scans in per day, but our employee increases to 3.4. Moving from 3 scans in per day to 3.4 is not necessarily enough to indicate outlier behavior, especially if this change is gradual. Still, the difference between this employee's behavior and their (former?) peer group's behavior is much more significant.

Common Classes of Technique

Now that we have an understanding of some of the causes of time series outliers, we will wrap up with a general classification of techniques. There are three common classes of time series analysis technique (Mehrotra et al., 69). The first type is *kernel-based techniques*, in which we compute some similarity measure across the entirety of the time series. From there, we could perform anomaly detection using classical techniques, such as calculating the similarity measure for a data point's nearest neighbors.

The second class of technique is *window-based techniques*. These techniques look at windows of time series data and compute an anomaly score for each window in the sequence. Then, we can combine the score of each relatively small window to calculate an overall anomaly score.

The third class of technique is *Markovian techniques*, patterned after *Markov chains*. A Markov chain is a state-based process in which the current value is highly dependent on the last value but any previous values have no significance. In other words, the value of x_t depends on x_{t-1} but not on x_{t-2} or earlier values. Markovian techniques tend to relax this requirement a bit, instead limiting us to some relatively small number of prior data points. An example of this is that the expected price of a stock today will depend heavily on what it was over the past few days but not at all on what it was six months ago.

We will primarily focus on window-based techniques throughout Part IV, as it gives us the clearest picture of which segments of time are most likely to be outliers. Kernel-based techniques are great if we have a large number of relatively short series and want to find the series-wide outliers; Markovian techniques, meanwhile, are great for tracking drastic, unexpected changes in short-term behavior within a series.

Conclusion

This chapter provided an overview of time series analysis. We defined what time series analysis is and learned what makes time series analysis different from non-time series analysis. From there, we reviewed several common time series analysis models. These models are primarily useful for predicting future time series values based on historical information, but it is possible to use the difference between a prediction and its actual value to provide an indication of whether a data point is an outlier. After covering time series analytical techniques, we gained a better understanding of what it means for a data point to be an outlier in a time series. Finally, we learned about three major classes of outlier detection technique for time series analysis.

In the next chapter, we will implement single-series time series outlier detection using one common technique: change point detection.

## CHAPTER 14

Change Point Detection

The prior chapter gave us an introduction to time series analysis, allowing us to gain an appreciation for how it differs from standard univariate or multivariate analysis. One of the biggest differences in time series analysis is that we often do not care about a single outlier point itself; instead, we care about changes in overall system behavior—in particular, unexpected changes in system behavior. This is where the concept of change point detection becomes important.

In this chapter, we will define change point detection and understand its benefits vs. "single-point" time series outlier detection. After that, we will look at two types of change point detection. Then, we will see how to implement change point detection using the ruptures package in Python. We will focus on two key techniques and then implement one of them in our outlier detection engine. Finally, like other code development chapters, we will update tests and integrate our changes into the website, making it easy for users to perform time series analysis.

What Is Change Point Detection?

Change point detection is the analysis of alterations in the patterns of time-variant signals. Returning to a theme from the prior chapter, we can break a time series down into four components: trend, seasonality, periodicity, and noise. We would expect a time series analysis based on constant fundamentals to stay fairly stagnant, in the sense that we may see factors such as seasonality or periodicity but that we continue to see these factors to approximately the same extent throughout our time series. What draws our attention in time series analysis is typically a change in the underlying fundamentals, which shifts the behavior of the time series itself. These shifts are also known as *change points* and allow us to separate our dataset into *segments* of data. For each segment, we assume that all data points follow essentially the same rules and are based on the same fundamentals. As we move between segments, some underlying mechanism has changed the fundamentals. Figure 14-1 shows an example of one artificial time series with four change points breaking data into five segments.

247

K. Feasel, *Finding Ghosts in Your Data*, https://doi.org/10.1007/978-1-4842-8870-2_14

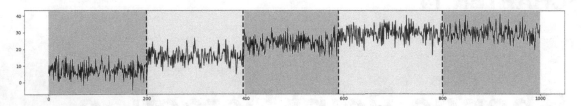

Figure 14-1. *One dataset with four change points*

Specifically, a change point is a change in one of three characteristics, which we will call level, angle, and size. In Figure 14-1, we see each change point affects the level—each segment is approximately flat, allowing for noise—but we do not see a change within segments in the direction of movement (the angle) or the difference between periodic highs and lows (the size). Our goal with change point detection is to track when there is a change, regardless of whether it is with respect to level, angle, or size.

Benefits of Change Point Detection

Up to this point, we have assumed that each data point in our analysis is independently distributed, meaning that prior data points' values do not affect our current data point. We learned in the prior chapter that time series analysis changes this due to autocorrelation: now, the prior value(s) should affect our current value. This leads to the first benefit of change point detection: if we find points that are substantially different from prior points, we have likely found something of interest to our audience. These change points are accordingly natural points of interest. Change points also help ensure that we capture the beginning of a new segment rather than every point in the segment, which reduces the number of data points to review.

Furthermore, change point detection can help with a common problem in time series analysis: dirty data. For example, suppose we wish to track the temperature of meat in a smoker using a thermometer. We might see the temperature slowly rising from 125 to 135 degrees Fahrenheit. Then, the thermometer suddenly registers a reading at 375 degrees. This data point is an outlier, and using time series techniques like we discussed in Chapter 13, we would mark this point as an outlier. With change point detection, however, we want to wait for a few more readings to determine if the temperature drops back down to 135 (give or take) or if it stays around 375—or even increases further. A temporary blip like this often indicates a faulty signal, error in reading, or error in transcribing a value rather than a change in the underlying fundamentals of our system—in this case, the process of barbecuing meats. When using change point detection, we will tend to filter out one-off blips if we quickly return to the original pattern.

Ultimately, the choice of whether to include one-off outliers is up to you, and we could combine change point detection with model-based time series outlier detection to get a more robust understanding of our data. For this chapter, however, we will continue our focus on change point detection.

Change Point Detection with ruptures

The ruptures library in Python is one of the most complete and popular libraries for change point detection. It comes with a series of built-in techniques for performing change point detection, making it easy for us to get started. Each technique combines three operations: a cost function, which we want to minimize; a search method, which helps us determine whether we have actually reached a change point; and a penalty function, which adds cost based on the number of change points we detect, making it costly to predict a large number of change points (Truong et al., 4–5). In this section, we will look at two of these techniques: dynamic programming and PELT.

Dynamic Programming

Dynamic programming is a technique that provides us an optimal solution, meaning that it will find the best solution given an input dataset. This comes with one unfortunate requirement: we need to know the number of *breakpoints* (i.e., the data points that separate segments) in our dataset before we can use dynamic programming. If we happen to know that answer, this technique will be very useful to us. In our case, however, we will not know the answer beforehand, as we make no assumption that the caller of our anomaly detector is aware of the distribution of their data. Asking them to provide this number to us would not be in line with our expectations.

Happily, there are other techniques that do not require foreknowledge of the number of segments, PELT being one of them.

PELT

Pruned Exact Linear Time (PELT) is another change point detection technique. Unlike dynamic programming, PELT is an approximate technique, meaning that it does not guarantee that it will find the best solution for a given input dataset. Instead, it will find something that is often good enough. There are two reasons why we prefer using a technique like PELT over dynamic programming. The first reason is the obvious one: dynamic programming requires us to send in the number of breakpoints, whereas PELT does not.

The other reason is a bit more prosaic: dynamic programming is a slow algorithm. Even if we were able to use it in our anomaly detection engine, it would be considerably slower for large datasets than PELT. Beyond this, ruptures offers an optimized version of PELT in its `KernelCPD()` function, which is written in the C programming language rather than Python in order to improve the function's performance even further.

Implementing Change Point Detection

Now that we have an idea of which library and technique we will use, let's start writing some code. First, we want to update `main.py`, the controller page for our FastAPI implementation. Listing 14-1 shows the implementation for time series analysis.

Listing 14-1. FastAPI's method for performing single time series analysis

```python
class Single_TimeSeries_Input(BaseModel):
    key: str
    dt: datetime.datetime
    value: float

@app.post("/detect/timeseries/single")
def post_time_series_single(
    input_data: List[Single_TimeSeries_Input],
    sensitivity_score: float = 50,
    max_fraction_anomalies: float = 1.0,
    debug: bool = False
):
    df = pd.DataFrame(i.__dict__ for i in input_data)
```

```
(df, weights, details) = single_timeseries.detect_single_timeseries(df,
sensitivity_score, max_fraction_anomalies)

results = { "anomalies": json.loads(df.to_json(orient='records', date_
format='iso')) }

if (debug):
    results.update({ "debug_weights": weights })
    results.update({ "debug_details": details })
return results
```

Each data point needs to send in three parameters: a key, a datetime, and a value. We will take a list of these data points, as well as sensitivity score and max fraction of anomalies, just as we do with univariate outlier detection. From there, we will call detect_single_timeseries() from the single_timeseries.py file. Listing 14-2 shows what this function looks like.

Listing 14-2. The function to detect change points in a single time series

```
import pandas as pd
import numpy as np
from pandas.core import base
import ruptures as rpt

def detect_single_timeseries(
    df,
    sensitivity_score,
    max_fraction_anomalies
):
    # Weights is here as a future-proofing measure.
    weights = { "time_series": 1.0 }

    # Ensure that everything is sorted by dt
    df = df.sort_values("dt", axis=0, ascending=True)

    num_data_points = df['value'].count()
    if ... # data checks elided for the sake of parsimony
    else:
```

```
(df_tested, tests_run, diagnostics) = run_tests(df)
(df_out, diag_outliers) = determine_outliers(df_tested, tests_run,
diagnostics["num_iterations"], sensitivity_score, max_fraction_
anomalies)
return (df_out, weights, { "message": "Result of single time series
statistical tests.", "Tests run": tests_run, "Test diagnostics":
diagnostics, "Outlier determination": diag_outliers})
```

The structure of this function is similar to what we built in the univariate and multivariate scenarios, in that we create a weights dictionary, run tests, determine outliers, and return the outliers as well as weights and other diagnostic details. One big difference is that we sort the values by dt ascending, ensuring that our data is in time series order. Because we expect there to be serial correlation, we want to make sure that "near" points are close to one another, regardless of how a user sent in the data.

The next thing we need to do is run tests. We will have one test to run, change point detection. Within that one test, however, we have quite a bit of choice. Specifically, when using the KernelCPD() (kernel change point detection) function in ruptures, we have two separate parameters we control: the choice of kernel and the cost of adding a new break point (also known as the penalty). There are three kernels available to us in ruptures, and each determines the cost function for our search. The first kernel available to us is a linear cost function, also known as an L2 cost function. With this cost function, we create linear segments and calculate the squared difference between the actual signal and our estimated mean. In other words, this represents the least squared deviation.

The second kernel is a Gaussian model, leveraging our kernel as *radial basis function (rbf)*. This technique is more complex than the linear technique and performs well enough on a variety of tasks. If we were to pick a single kernel with no information about the underlying data, the Gaussian kernel would be the best starting point. Of course, additional information could lead us in the direction of one of the other kernels, so this is not guaranteed to be the best choice.

The third and final available kernel uses a technique known as *cosine similarity*. Cosine similarity is most useful in text segmentation, such as in the popular word2vec family of algorithms.

Each of these kernels, by way of implementing different cost functions, will necessarily give us different results. We cannot assume that our users will know which is correct—and we likely will not know the answer to that either—so we take the easy way out: build an ensemble and see what shakes out.

Listing 14-3 shows the run_tests() function. This function accepts a DataFrame and tries out each of the three kernel techniques for a variety of penalty factors ranging over 7 orders of magnitude. Penalty factors can make a big difference in our results, but there is no "most appropriate" penalty value across all datasets. Therefore, we want to capture results for a wide variety of penalty values and allow users to scale the numbers of results up and down using the sensitivity score and max fraction of anomalies thresholds as they do for univariate anomaly detection.

Listing 14-3. The function to run change point detection tests as an ensemble

```
def run_tests(df):
    tests_run = {
        "changepoint": 1
    }

    num_records = df['key'].shape[0]
    diagnostics = {
        "Number of records": num_records
    }
    signal = df['value'].to_numpy()

    kernels = { "linear", "rbf", "cosine" }
    penalties = { 0.001, 0.005, 0.01, 0.05, 0.1, 0.5, 1, 5, 10, 20, 50, 80,
    100, 200, 500, 800, 1000 }
    diagnostics["kernels"] = kernels
    diagnostics["penalties"] = penalties
    diagnostics["num_iterations"] = len(kernels) * len(penalties)

    scores = np.zeros([num_records])
    for idx,k in enumerate(kernels):
        algo = rpt.KernelCPD(kernel=k).fit(signal)
        for idxp,p in enumerate(penalties):
            result = algo.predict(pen=p)
            for ix,r in enumerate(result[:-1]):
                scores[r] += 1

    df["anomaly_score"] = scores
    return (df, tests_run, diagnostics)
```

In the preceding listing, we can see that there are three kernels and 17 choices for penalty weight, giving us a total of 51 separate executions of kernel change point detection. The way we will score change points in our input dataset is simple. First, we start with a zeroed-out array whose length matches that of our dataset. Then, for each kernel and penalty combination, we run kernel change point detection against the values in our dataset. When running predict() on our dataset, we get back an array that contains one value for each index that the algorithm marks as a change point, *plus* one value that represents the length of the incoming dataset. We do not need that final value; instead, for each change point, we increment the current score for that index in our scores array. At the end, each array index will have a score ranging from 0 to 51, with 0 meaning that no models determined this point was a change point and 51 indicating that every model found it as a change point. In practice, even extreme change points will end up with scores in the 10–20 range due to the way the penalty function interacts with any given dataset.

After scoring each data point, we need to determine which change points we want to mark as outliers. Listing 14-4 provides the code to do this.

Listing 14-4. The function to determine outliers in our DataFrame

```python
def determine_outliers(
    df,
    tests_run,
    num_iterations,
    sensitivity_score,
    max_fraction_anomalies
):
    sensitivity_threshold = (num_iterations / 1.5) * ((100.0 - sensitivity_
    score) / 100.0)
    diagnostics = { "Sensitivity threshold": sensitivity_threshold }

    max_fraction_anomaly_score = np.quantile(df['anomaly_score'], 1.0 -
    max_fraction_anomalies)
    diagnostics["Max fraction anomaly score"] = max_fraction_anomaly_score
    if max_fraction_anomaly_score > sensitivity_threshold and max_fraction_
    anomalies < 1.0:
        sensitivity_threshold = max_fraction_anomaly_score
```

```
    diagnostics["Sensitivity score"] = sensitivity_threshold
  return (df.assign(is_anomaly=df['anomaly_score'] > sensitivity_
  threshold), diagnostics)
```

The determination function looks similar to what we have for univariate and multivariate outlier detection. We first set a sensitivity threshold by scaling the number of iterations (i.e., the number of kernels times the number of possible penalty values) to increase the likelihood of data points appearing for a given sensitivity score. Then, we factor in the sensitivity score, ensuring that higher sensitivity scores lead to a lower sensitivity threshold and therefore more outliers. In the multivariate outlier detection scenario, sensitivity score became a very weak lever, to the point that users would not even need to think about it. With single time series, sensitivity score becomes relevant once again.

In addition to sensitivity score, max fraction of anomalies is still around, and it behaves the same as it did for the univariate and multivariate outlier detection scenarios, providing a maximum cap on how many data points we might mark as change points. The end result of this is an API endpoint whose behavior is very similar to univariate outlier detection. This will make testing and website implementation straightforward.

Test and Website Updates

As with our other endpoints, we want to ensure that this new endpoint behaves in a reasonable manner. To that end, we will incorporate unit and API tests into the mix. Additionally, we will need to think about how to display our results.

Unit Tests

There are three things we want to ensure with our single time series unit tests. First, we want to make sure that the code we wrote functions without error. Second, we want to make sure that the "direction" of the sensitivity score and max fraction of anomalies settings are both correct—in other words, we want a higher sensitivity score to give us more rather than fewer outliers. Finally, we want to see that the spread of outlier values is reasonable—ideally, we have a fairly smooth transition as we move from low sensitivity score like 5 up to our high of 100. Listing 14-5 shows an example of a test that happens to fulfill all three purposes for sensitivity score.

Listing 14-5. A unit test that covers a variety of sensitivity scores

```
@pytest.mark.parametrize("df_input, sensitivity_score, number_of_
anomalies", [
    (sample_input, 100, 8),
    (sample_input, 90, 7),
    (sample_input, 80, 7),
    (sample_input, 70, 4),
    (sample_input, 60, 3),
    (sample_input, 50, 1),
    (sample_input, 40, 1),
    (sample_input, 25, 0),
    (sample_input, 5, 0),
    (sample_input, 1, 0),
    (sample_input, 0, 0),
])
def test_detect_single_timeseries_sample_sensitivity(df_input, sensitivity_
score, number_of_anomalies):
    # Arrange
    df = pd.DataFrame(df_input, columns=["key", "dt", "value"])
    max_fraction_anomalies = 1.0
    # Act
    (df_out, weights, diagnostics) = detect_single_timeseries(df,
    sensitivity_score, max_fraction_anomalies)
    print(df_out.sort_values(by=['dt']))
    # Assert
    assert(number_of_anomalies == df_out[df_out['is_anomaly'] == True].
    shape[0])
```

This test covers 11 separate test cases, spanning sensitivity scores from 0 to 100. The sample input set is an artificially generated time series that contains 17 data points. The first 15 data points range from a low of 13.7 to a high of 16.5. The final two data points are 190.8 and 193.7, respectively, indicating that a major change has occurred. Although we should not draw wide-sweeping conclusions from a single dataset, we can see that there is some variety in our output starting at a sensitivity score of 40. We could surmise that a sensitivity score somewhere between 50 and 70 is probably the most reasonable choice for this dataset, as a review of the sample input (available in the accompanying code repository) shows that there is a rising action through the first six data points, a steady decline from the 7th through 10th data points, a flattening out of the decline from the 11th through 15th data points, and an order of magnitude jump with the 16th data point. Figure 14-2 shows a simple plot of this data, excluding the two largest outliers.

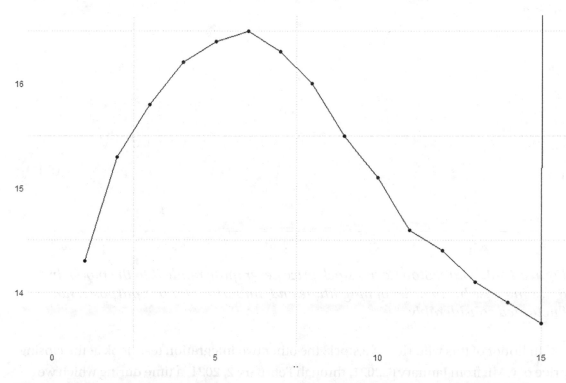

Figure 14-2. *Although these values range from 13.7 to 16.5, we can see definite change points in the data. The nearly vertical line at the end represents the jump from 13.7 to 190.8*

Integration Tests

Aside from unit tests, we will also want to create integration tests to ensure that API functionality works as we would expect. There are three new tests in the Postman collection for Chapter 14. One test covers a basic scenario in which there are no change points detected. Although there is motion in the data, no movement is sufficient to indicate to the outlier detection engine that there is a shift in the fundamentals.

The other two cases cover one of the most famous anomalies in recent memory: the meteoric rise and fall (and rise and fall and rise…) of GameStop Corp's stock, traded as

GME on the New York Stock Exchange. The story of how a stock that traded anywhere from $4 to $20 per share from 2017 through January 12, 2021 could double one week later and hit a peak of $347.51 is an amazing story, one this book is ill-equipped to tell. Nonetheless, from Figure 14-3 alone, we can see that the share price for GameStop Corp has fluctuated wildly over the past 18 months from January 2021 through the time of writing.

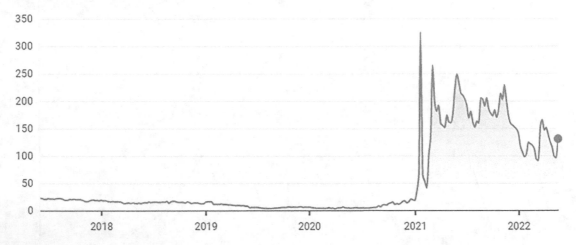

Figure 14-3. *GameStop Corp's stock price never quite made it to the moon, but it did increase by an order of magnitude and, important for our purposes, has fluctuated considerably*

In honor of this wild ride of a stock, the other two integration tests look at the closing price of GME from January 6, 2021, through February 2, 2021, a time during which we note five separate change points with a sensitivity score of 60.

Website Updates

The final major consideration for this chapter is to update the website to support single time series analysis. In `site.py`, we will need to make a few changes. First, we will include a default dataset that can show new users how to structure their data for this single time series call. Like with the integration tests, we will use GME stock data from January 6, 2021, through February 2, 2021, as our example. We will also create a function to convert a single time series list into a JSON array. Listing 14-6 shows this block of code, which is very similar to the multivariate conversion function.

Listing 14-6. Convert an array of time series data to JSON for use in our anomaly detection engine.

```
@st.cache
def convert_single_time_series_list_to_json(time_series_str):
    mv_ast = ast.literal_eval(time_series_str)
    return json.dumps([{"key": k, "dt":dt, "value": v} for idx,[k,dt,v] in
    enumerate(mv_ast)])
```

The other significant change concerns how we will display our time series data. Typically, line charts are a great choice for displaying time series data, especially when the time series displays factors like seasonality or periodicity. In order for us to show outliers as points while still seeing the entire trend as a line, we will use two separate charts. Listing 14-7 shows the code we will use to display these charts.

Listing 14-7. Display a line chart as well as a scatter plot.

```
elif method=="timeseries/single":
    st.header('Anomaly score per data point')
    colors = {True: '#481567', False: '#3CBB75'}
    l = px.line(df, x=df["dt"], y=df["value"], markers=False)
    l.update_traces(line=dict(color = 'rgba(50,50,50,0.2)'))
    s = px.scatter(df, x=df["dt"], y=df["value"], color=df["is_anomaly"],
    color_discrete_map=colors,
                symbol=df["is_anomaly"], symbol_sequence=['square', 'circle'],
                hover_data=["anomaly_score"])
    g = go.Figure(data=l.data + s.data)
    st.plotly_chart(g, use_container_width=True)

    tbl = df[['key', 'dt', 'value', 'anomaly_score', 'is_anomaly']]
    st.write(tbl)
```

The line chart includes all data points, and we use the update_traces() method to change its color to a neutral gray, making it fairly easy to see while not overwhelming the actual data points, which we generate from a scatter plot. Finally, we use Plotly's graph_ objects to create one figure from two, calling go.Figure() and passing in both plots. Figure 14-4 shows an example of this output.

259

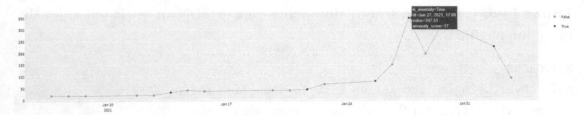

Figure 14-4. *A visual representation of change points in GME stock in January of 2021*

When mousing over a single data point, we can see not only whether it is an anomaly but also the anomaly score—that is, how many individual tests marked this data point as a possible change point. The debugging information for this method will also show the minimum threshold for what constitutes an outlier. For example, using a sensitivity score of 65 and a max fraction of anomalies of 0.26, the minimum anomaly score that would trigger the system is 13, meaning that the value on January 19 is not an anomaly as only 12 of the tests picked it as a change point. January 21, however, does show up as a change point because it had 14 of the 51 possible tests flag it as a change point.

Avenues of Further Improvement

Before closing this chapter, it is worth spending a small amount of effort thinking about ways to improve our single time series anomaly detection engine. The most obvious answer would be to incorporate point-based outlier detection as well as change point detection. In the prior chapter, we looked at modeling techniques like ARIMA that allow us to generate predictions of where we believe the data ought to be based on historical information. Using this process, we could create a *backcast*, that is, a prediction of events that have already occurred to see how our forecasting model would have behaved given available information. From there, we could use distance measures like we developed for univariate outlier detection to adjudge the likelihood of a given actual data point being an outlier based on what we expected given our model.

The major sticking point with this approach is that we would need a solid model, which implies a certain understanding of the underlying fundamentals driving the time series. For example, modeling techniques like ARIMA generally expect you to understand the periodicity of your data, knowing for example that a cycle occurs every 12 data points. In order to implement this in a general-purpose outlier detection engine, we would need to pass that requirement on to the end user, who likely does not know the correct answer. Furthermore, there are quite a few modeling techniques for time series analysis, and—similar to what we saw with kernel change point detection before—the results may differ radically between them. We could potentially try several models—including different combinations of hyperparameters for techniques like ARIMA—and determine which fits most closely to the resulting stream of data. Then we could choose that best-fit model as our candidate and calculate distances from it. I leave this as an exercise to the reader, should you wish to give it a try.

Conclusion

In this chapter, we looked at one way of performing outlier detection on a single time series: tracking each change point. Although change points are not necessarily anomalous—they may very well be embedded into the underlying logic of the phenomenon we wish to investigate—they are a good starting point for investigation. In addition to change point detection, we could incorporate other techniques such as calculating the distance from an expected model.

In the next chapter, we will generalize the problem of single time series analysis and ask the question, what if we want to analyze more than one series at a time?

## CHAPTER 15

An Introduction to Multi-series Anomaly Detection

Chapters 13 and 14 dealt with one form of time series anomaly detection: the single time series. With that understanding in place, we will extend our focus to situations in which there are multiple series of data that we want to analyze in conjunction rather than independently. In this chapter, we will formalize our understanding of multi-series time series, focusing on how it differs from the single series approach we took earlier. Then, we will look at four common techniques for analyzing multi-series data. Before we wrap up this chapter, we will review some common problems when trying to analyze multi-series data before we implement one algorithm in Chapter 16.

What Is Multi-series Time Series?

Our analysis so far has focused on a single time series—that is, one line on a chart. Our key technique has been to use change point detection to see when that line alters significantly. By contrast, multi-series time series involves multiple phenomena that are related to one another in some fashion, but unlike univariate vs. multivariate anomaly detection, multi-series is not the same as single series with multiple non-time input variables. Instead, multi-series time series analysis will still feature univariate data (plus time) but will feature it for multiple, related series of data. For example, Figure 15-1 shows the Earned Run Average of each team in Major League Baseball's National League East between the division realignment of 1994 through 2019. We could potentially analyze this as a multi-series time series problem because there is (potentially!) enough similarity between the series to make them interesting: we are looking at the same sport,

K. Feasel, *Finding Ghosts in Your Data*, https://doi.org/10.1007/978-1-4842-8870-2_15

the same division, the same variable, and over the same time frame. We might assume that the rules of the game and league-wide trends (such as around use of performance-enhancing drugs) will drive a fair amount of behavior across teams.

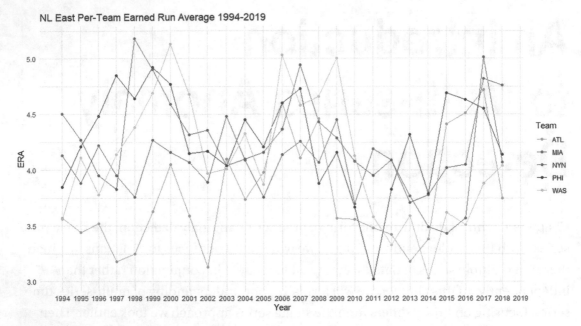

Figure 15-1. *Earned Run Averages across the Eastern division of the National League, 1994–2019. For the sake of simplicity, the Montreal Expos were coded as WAS and Florida Marlins as MIA*

Most likely, you see a jumbled mess in the preceding figure. With enough patience, we might be able to make out general trends from this analysis, but it is hard to separate out the "micro" effects of team behavior from the "macro" effects of league-wide (or at least division-wide) changes over time due to rules and team behaviors using just our eyes. This is one of the primary differences between single-series and multi-series analysis, but it is by no means the only one.

Key Aspects of Multi-series Time Series

There are several things to keep in mind when looking at multi-series time series analysis. First and foremost, our goal is to perform some sort of functional comparison of the shape of multiple series of data. We want to understand how the behavior in one set affects, is affected by, and moves as the result of common causes with other sets of data. To do this, we need to focus on overlapping time frames. This does not necessarily mean that all series need to have identical points of time. For example, we might track one measure every three seconds and another measure every five seconds. Even though the two sets do not overlap exactly, as long as we can interpolate values with reasonable accuracy, we could compare these two datasets. Most often, to solve a time frame mismatch problem, we can use a technique known as *linear interpolation*, drawing a straight line from the prior data point to the next and estimating the value at any point in between known points based on the line we've created. We do this because we do not have sufficient information to know the exact behavior of the series, and so, instead of trying to fit some hypothetical curve, we draw the simplest line and call it a day.

Beyond this, we might want to normalize our datasets, ensuring that data points range between 0 and 1, inclusive. This can help us track shifts in series regardless of their base values or typical variance. As an example, suppose we measure temperature changes in tenths of a degree and efficiency ratings in percentage points, typically ranging from 85 to 95. Plotting both of these measures on the same graph would make it difficult to determine if the two attributes are moving in similar patterns. If we normalize each series to [0, 1], it is easier to spot patterns in the data. Figure 15-2 is the same Earned Run Average dataset for National League East teams from 1994 through 2019, this time normalized such that each team's lowest seasonal ERA is 0 and the highest ERA for each team is 1.

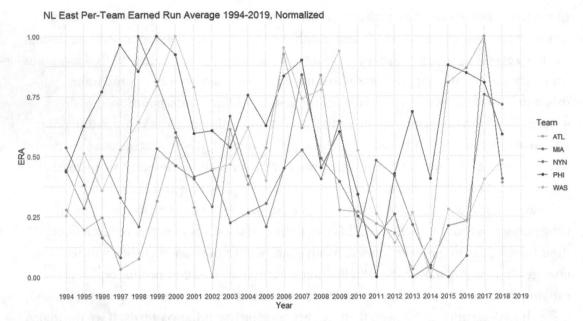

Figure 15-2. *Normalized Earned Run Averages across the Eastern division of the National League, 1994–2019*

By normalizing the dataset, we can see a small pattern in which the teams of the NL East pitched markedly better as a whole from approximately 2010 to 2014 (because lower values of Earned Run Average are better) and they all fared relatively poorly from 2006 to 2009 and 2017 to 2019.

Along with normalizing data, we may also decide to *censor* it, particularly because our aim is to find outliers and normalization tends to smother outliers, as we learned in the univariate and multivariate cases. One common approach for censoring comes from Mehrotra, in which you perform normalization on data points between the 10th and 90th percentiles, allowing values below the 10th percentile to become negative and values above the 90th percentile to go above 1.0 (Mehrotra, 102). By doing so, you can still observe trends without one extremely large or extremely small data point skewing the rest of the time series. This is very similar to the technique we used with Box-Cox transformation in Chapter 7.

What Needs to Change?

As we shift our thinking from single series to multi-series, we should keep the following differences in mind. Ultimately, these are considerations as we begin to think in two directions: forward in time and up and down between series.

What's the Difference?

One important question we want to ask is, what is the difference we want to track? We have two major categories of difference between series: distances of points and differences in direction. In the former case, we would be concerned about a data point whose value differs significantly from its peers at the same point in time. If one team consistently has a higher (or lower) ERA than other teams during the time frame, we would naturally consider this difference the most anomalous when the rest of the division is at its furthest point from the team in question. A good example of this is the Washington Nationals in 2017. Their ERA had increased along with the rest of the division, but it was significantly better than all of the other teams in the division. A similar example is the Atlanta Braves in 2002, following a relative downswing in ERAs in the division but taking it further than the other teams.

By contrast to distance measures, we could also look at differences in direction: if one series zigs while the others zag, this could be of interest, particularly if the magnitude of this change is significant and we have reason to believe that the underlying fundamentals should be similar for each series. If average home sales prices in markets are generally trending up and one city is moving in the opposite direction, that case now becomes a curiosity.

Beyond this "micro" view of comparing one series vs. others, we have a third concept of difference: changes in the aggregate. In this case, we care less about how any individual series moves but more about how all series as a whole move. If the overall trend among series in our dataset changes, it may reflect a change in the fundamentals that could be of interest to us. Going back to our baseball example, there have been several rules and behavioral changes over the past 150 years that have affected measures like Earned Run Average: the composition of the ball, how frequently umpires switch out balls, the height of the pitching mound, the distance from the mound to home plate, how umpires interpret the strike zone, when and how frequently to switch out pitchers during the game, off-season training regimens, and much more have affected the relationship between pitcher and batter over the years. With a major rule change, such as increasing the height of the pitching mound or moving it closer to home plate, we would expect to see a general movement toward lower Earned Run Averages as pitchers gain an advantage relative to batters. Similarly, rule changes like a tighter strike zone or introducing a designated hitter to bat in the pitcher's stead would lead us to anticipate higher ERAs.

Leading and Lagging Factors

Another thing to keep in mind is whether we expect any lags in how one series affects another. For example, suppose we keep track of two series: one that indicates the temperature control setting on a thermostat and the other that indicates the current temperature of a room. Assuming that the temperature control setting is configured correctly (which may not always be the case in offices or retail stores) and is always running, a change in the temperature control setting will necessarily precede a corresponding change in room temperature, steadily increasing or decreasing temperature until we reach an equilibrium state in which either the current room temperature matches the control setting or the heating and cooling unit is incapable of changing the temperature any further given outside conditions. Therefore, if we track these two series as is, we might notice inconsistent behaviors: a person increased the temperature in the control unit, yet the temperature continues to decrease. This is because the change is not instantaneous and the air conditioner may continue to run for a couple of minutes before the unit's controller sends signals to stop condensing and blowing air. If we shift the time of temperature observation forward several minutes, we should see more consistent behavior with respect to thermometer settings.

Available Processes

There are two classes of approach for solving multi-series time series outlier detection problems. The first approach is the model-based approach we discussed in Chapter 13: if we build a model using the ARIMA family, exponential smoothing, or some other technique, we can generate expectations based on the existing fundamentals. The authors of these approaches, however, intended us to use them for analyzing a single series, not comparing multiple series. For this reason, attempting to implement a technique like ARIMA for multi-series comparison is clunky (to say the least).

Instead, we tend to approach multi-series time series analysis from what Mehrotra et al. call the "transformation approach" (Mehrotra, 172–173). The key insight of this approach is that we typically do not want to deal with each series as a whole series. Instead, we can perform three operations: *aggregation, discretization,* and *signal processing.* The first operation, aggregation, compacts together different values in the time series based on the assumption that movement within the series will be smooth. By reducing the number of data points in a given series, it makes analysis easier to perform. It also gives us a convenient method for ensuring that each data point for each series matches up in time. For example, if we capture a signal from one piece of equipment every minute and another piece of equipment every 28 seconds over the course of a month, we could aggregate both of them to an hourly interval. This drastically reduces the number of data points we need to consider for outlier detection and also reduces the chance of noise in a series triggering whatever algorithm we choose to implement. Within each hour-long block, it is up to us to determine how we aggregate these results.

Along with aggregation, we can incorporate the second operation, discretization. The idea behind discretization is that we apply a single value to the series as a whole, based on its general characteristics. For example, we might discretize each series into one of three values: rising, falling, and steady. This way, we only need to keep track of one piece of information rather than the combination of values. In doing so, we lose potentially useful information that could add nuance to our analysis, but it does allow us to solve complex problems with much less computational effort. We can combine this with aggregation fairly easily: the aggregation process breaks each series into a fixed-size window, and the discretization operation allows us to describe each fixed-size window based on its general characteristics.

Another technique is to perform signal processing, typically using a Fourier or Wavelet transformation. A full discussion of this is outside the scope of this book, but the treatment of time series as a signal is a powerful technique and the one that drives the ruptures library we used in the prior chapter.

Now that we have reviewed some of the key techniques for series transformation, let us move to three common algorithms we can use to discretize our aggregated series. Each algorithm assumes we want to compare segments from two time series, but in the next chapter, we will look at a way to generalize this process.

Cross-Euclidean Distance

The first technique is called Cross-Euclidean Distance. Listing 15-1 shows the formula for calculating the Cross-Euclidean Distance for two time series.

Listing 15-1. The formula to calculate Cross-Euclidean Distance for two time series

$$d_E = \sqrt{\left(\sum_{t=1}^{n}\big(x(t)-y(t)\big)^2\right)}$$

Explaining the formula, we calculate for each point in time the distance between the data points for the two series at that time, sum the squares of those distances, and take the square root of the result. To provide an example, suppose that our X series includes the Earned Run Averages for the Atlanta Braves from 1994 through 2002: X = { 3.57, 3.44, 3.52, 3.18, 3.25, 3.63, 4.05, 3.59, 3.13 }. Our Y series includes the Earned Run Averages for the New York Mets during the same time frame: Y = { 4.13, 3.88, 4.22, 3.95, 3.76, 4.27, 4.16, 4.07, 3.89 }. The first step of the calculation is to square the differences between our Atlanta and New York data points, giving us S = { 0.31, 0.19, 0.49, 0.59, 0.26, 0.41, 0.01, 0.23, 0.58 }. From there, we sum the squares and take the square root of the result, giving us a Cross-Euclidean Distance of 1.75. We could compare this between teams—assuming that we keep the number of elements we compare the same—and know that a higher Cross-Euclidean Distance indicates a bigger difference between the two series.

Cross-Correlation Coefficient

Cross-Euclidean Distance is a useful technique, but it has the downside that it is calculated on the same scale as our underlying feature. With something like Earned Run Average, this can make a big difference for two reasons: first, the average team ERA changes over time, and those changes can be fairly drastic. For example, the league-wide ERA from 1914 to 1918 was approximately 2.75–2.80. Moving 80 years into the future, the league-wide ERA from 1994 to 1998 was approximately 4.50. A Cross-Euclidean Distance of 1.75 in 1990s terms can be quite different from one in 1910s terms. Furthermore, the way Earned Run Average is calculated, there is a hard floor at 0.00 and in practical terms, a floor at 2.00—in fact, from 1910 through 2020, only the 1910 Philadelphia Athletics had a team ERA below 2.00; from 1920 forward, the lowest team ERA was the 1967 Chicago White Sox, which had a team ERA of 2.45. This tells us that when the league-wide average is nearer to 3.00–3.20, we are liable to have a smaller range of values than an era in which the team ERA is 4.50 or higher. For this reason, we might wish to use *Cross-Correlation Coefficient*, which we determine using the formula in Listing 15-2.

Listing 15-2. The Cross-Correlation Coefficient gives us a scale-invariant comparison between two time series segments

$$d_C = \sqrt{\left(2\left(1 - corr\left(X,Y\right)\right)\right)}$$

In this case, `corr` represents the Pearson correlation coefficient between the time series segments. In the Atlanta-New York case before, the two teams' pitching staffs have a Pearson correlation coefficient of 0.26, indicating that there was a minor-to-moderate positive correlation in ERA between the two staffs. Performing the inner calculation, we have 2 * (1 – 0.26) = 1.48. Taking the square root of this value gives us a Cross-Correlation Coefficient of 1.22. Note that this number is *not* directly comparable to Cross-Euclidean Distance, but it is directly comparable to other Cross-Correlation Coefficients, such as if we performed the same exercise for two franchises in the 1910s instead of the 1990s.

SameTrend (STREND)

The final technique we will cover in this chapter is called SameTrend, otherwise known as STREND. With STREND, we follow a three-step process. The first step is to calculate differences in the X and Y series. Continuing our example from before, we have nine data points for Atlanta and the corresponding nine data points for their divisional counterparts in New York. Calculating differences means that we take the value at time t and subtract it from the value at time t+1. For example, the first two values of ERA for Atlanta are 3.57 and 3.44, meaning that the first difference calculation will be -0.13. In total, we will get eight difference calculations, as we have nine data points. For Atlanta, these are { -0.13, 0.08, -0.34, 0.07, 0.38, 0.42, -0.46, -0.46 }, and for New York, they are { -0.25, 0.34, -0.27, -0.19, 0.51, -0.11, -0.09, -0.18 }. These form Δx and Δy, respectively.

Listing 15-3 shows the next step, which calculates the value of S at some given time t. The trick with STREND is that we do not care about the *magnitude* of changes; we only care about the *direction* of changes and specifically, if both series move in the same direction or not.

Listing 15-3. The second step of STREND is to calculate S(t)

$$S(t) = \begin{cases} 1\ if\ \Delta x(t) \cdot \Delta y(t) > 0 \\ -1\ if\ \Delta x(t) \cdot \Delta y(t) < 0 \\ 0\ otherwise \end{cases}$$

Calculating S(t) for our pitching staff set, we get a result of { 1, 1, 1, -1, 1, -1, 1, 1 }. If there were cases in which either team's ERA did not change between seasons, we would have marked that with a 0, but we do see movement in each season. We finish our calculation of STREND using the formula in Listing 15-4.

Listing 15-4. The formula to calculate the SameTrend value from two time series segments whose movement is defined by some S(t)

$$d_S(X,Y) = 1 - \frac{1}{n-1} \sum_{t \in [1..n-1]} S(t)$$

Summarizing S(t), we end up with a value of 4. Plugging this into our equation with the knowledge that n is 9, we end up with a SameTrend value of 1 − 4/8 = 0.5. With SameTrend, lower numbers indicate that the two series behave in a similar fashion, with 0 being a perfect correlation between the two, 1.0 indicating absolutely no correlation, and values above 1.0 indicating negative correlation. The highest possible value is 2.0, which would indicate perfect negative correlation.

Common Problems

Before wrapping up this chapter, we will look at three common problems we may run into when performing multi-series time series outlier detection. The first problem comes when we attempt to apply modeling techniques to time series data. In that case, we need to bear in mind that each time series modeling approach includes an error term, indicating a range of values that would not violate the model. To that extent, suppose one series trended somewhat upward but the other series trended somewhat downward. Suppose further that both series are still within range of their respective error terms.

Even though one has moved up and the other down, our model is not strong enough to indicate that there actually is a true difference in trends. Our eyes may be able to spot something, but we have not escaped "normal" behavior for the model. Using a modeling-based approach may cause us to fail to find valid behavioral differences.

The second problem we will likely run into happens when using the transformation approach. In this case, we have numbers that represent distances on some scale. We are able to say that one range is different from another and that the two series are more different at some point than at another point. What we are not able to say, however, is when a value for SameTrend, Cross-Correlation Coefficient, or Cross-Euclidean Distance is sufficiently large as to merit marking the segment as an outlier. Like our situation with multivariate outlier detection techniques such as LOF and COF, we will need to determine some "good enough" measure.

The final problem of importance here is that these techniques are all built around comparing two series of data. As soon as we introduce a third series, we have a new realm of complexity. We could conceivably perform pairwise comparisons of each series, but with 20 separate series, this could turn into a logistical nightmare. In the next chapter, we will adroitly avoid the first problem by not using a model-based solution and take on the second and third problems instead.

Conclusion

In this chapter, we gained some understanding of how multi-series time series analysis differs from single-series analysis. We considered some of the technical difficulties of analysis, such as one series being a lagging indicator of another. After that, we looked at three techniques for discretizing time series segments, allowing us to compare two time series as a whole without loss of significant information. Finally, we ended with a discussion of some common problems in multi-series time series analysis.

In the next chapter, we will introduce one more technique for discretizing data, implement this technique in our code base, and tackle the problem of multi-series time series outlier detection in Python.

CHAPTER 16

Standard Deviation of Differences (DIFFSTD)

In the prior chapter, we learned about several pairwise techniques we can use to perform multi-series time series analysis. In this chapter, we will introduce one more such technique and will use it as the first method in a multi-series time series ensemble.

What Is DIFFSTD?

Standard Deviation of Differences (DIFFSTD) is a pairwise time series comparison technique, meaning that we can use it to compare behavior between two series and gain an understanding of how data movements in those two series diverge over time. As we learned in the prior chapter, there are two operations we should perform for multi-series time series analysis: discretization and aggregation. Discretization will allow us to generate smaller segments of data, which we can then aggregate and compare using techniques like DIFFSTD. In this section, we will assume that we have already discretized the data so we can focus on how to calculate DIFFSTD; when we begin to write code to implement the function, we will need to tackle the problem of discretization as well as aggregation.

Calculating DIFFSTD

Calculating DIFFSTD is fairly straightforward. Listing 16-1 shows the formula for comparing one segment for each of two series.

© Kevin Feasel 2022
K. Feasel, *Finding Ghosts in Your Data*, https://doi.org/10.1007/978-1-4842-8870-2_16

Listing 16-1. The formula to calculate DIFFSTD

$$dist(X,Y) = \sqrt{\frac{\sum_t (\delta(t) - \mu)^2}{n}}$$

In this formula, X and Y represent matching segments of two separate time series. The value $\delta(t)$ represents the numeric difference between X and Y at time t. The value μ represents the mean of the difference between X and Y for all n data points we see in the segment.

For example, suppose we have a segment X whose values over the relevant time frame are { `15.3, 15.8, 16.2, 16.4, 16.5` }. We also have a segment Y whose values over the same time frame are { `26.5, 26.3, 26.0, 25.5, 25.1` }. The differences between X and Y in this set are { `-11.2, -10.5, -9.8, -9.1, -8.6` }. The mean of these differences is -9.84. Knowing these values, we can subtract the mean from each value of $\delta(t)$. For example, for the first pair of data points, `-11.2 - -9.84 = -1.36`. We repeat this for each subsequent data point, coming up with the set { `-1.36, -0.66, 0.04, 0.74, 1.24` }. The next step is to square each of these results, giving us { `1.8496, 0.4356, 0.0016, 0.5476, 1.5376` }. The sum of these five data points is 4.372, so if we divide it by the number of data points we have (which is 5), we get the value 0.8744. Finally, the square root of this is 0.935, giving us our result.

In order to interpret this result, we first need to note that all DIFFSTD results will be greater than or equal to 0. If the result is equal to 0, it means that the two segments do not diverge in behavior at all: when one moves up, the other moves up in exactly the same way; when one moves down, the other moves down in exactly the same way. Any result above 0 indicates some amount of difference in behavior. Our example is of two datasets moving in opposite directions: one is slowly trending upward and the other slowly trending downward. This causes our DIFFSTD value to be above 0, but because those trends are small, our DIFFSTD is fairly small as a result.

Key Assumptions

Before we move on to writing code, we should lay out a few key assumptions for how multi-series time series analysis will work. These assumptions are simplifying assumptions but are not, strictly speaking, absolutely necessary for multi-series analysis.

The first assumption is that each input time series will be of the same length—that is, every series will have the same number of values. Furthermore, the element at point t in each series will reflect the same point in time, regardless of series. For example, if we are comparing daily stock performance data for a pair of stocks, we will require that the two series have the same number of data points and that those points represent the same dates. We could relax this assumption with additional development effort by creating synthetic time series that map from the two (or more) input series to a set of new series with common time elements. For example, if one series is missing a data point for some particular date, we might interpolate the value for that date based on the surrounding data. We can do this because we assume that each time series exhibits autocorrelation and so demonstrates some amount of stability between data points.

The second assumption we will follow is that all of the time series we receive as inputs should reflect the same underlying phenomenon. For example, if we receive a time series representing the number of suits sold to men and a time series representing the average salary of a professional basketball player, we assume that there is some underlying force or common set of forces driving the two series. This assumption is important because our aim is to find points of significant divergence. If there is no underlying factor that drives the two series, we would naturally expect divergence. We could still map the areas of increased divergence, but there is nothing for a person to learn or do as a result. This is an assumption we will not be able to relax, as we would not be able to write code to "ensure" that the time series do bear some relationship with one another.

The final assumption we have is that there is no lagged autocorrelation between the series, or at least that lagged autocorrelation is not a significant factor. In the prior chapter, we covered an example of lagged autocorrelation: the temperature control setting on a thermostat vs. the temperature of a room. In this case, we do have knowledge that these reflect the same underlying phenomenon: the temperature control sensor manages the heating and cooling system that regulates room temperature, driving room temperature toward the thermostat's setting. We would expect there to be lags, however, as the temperature in the room does not drop several degrees simply because somebody changes the thermostat setting. Instead, changing the thermostat setting may prompt the temperature control system to engage, blowing warmer or colder air into the room until its temperature sensor indicates that the room has reached the preferred temperature. If we were to analyze the time series of thermostat settings vs. the time series of room temperatures, we might end up with incorrect results because we did not take lagged autocorrelation into account. On a hot day, the temperature in the room may continue to rise even though a person has lowered the thermostat control setting's temperature value, as it may take the equipment several minutes to condense and blow enough air into the room to begin cooling it down. If we know what that time difference is (or at least have an understanding of how long it normally is), we could shift the room temperature series left by that amount and control for the lag, giving us better results. For our general-purpose outlier detection engine, we will not know details about the lag, and for the sake of simplicity, we will assume there is none.

If you did wish to introduce a feature that tracks lags in datasets, the end user would need to know more details about the distribution of their data. Assuming this is the case, and—without loss of generality—we say that series Y lags behind series X by two steps, we could shift series Y left by two steps so that what we measured at time t would be represented at time t-2, as this is when the change in X prompted the change in Y. Doing this would leave two data points for Y at the beginning of the series with no corresponding X value and two data points at the end of the series with no corresponding Y value. We would then remove those mismatched beginning and end points and analyze the remainder. Another solution to this problem would be to have users perform the de-lagging themselves, assuming they have an understanding of the lags.

Now that we have covered the three key assumptions we will make for this chapter, let's start writing some code.

Writing DIFFSTD

In this section, we will put together a solution for multi-series time series analysis. Because there is no readily-accessible library that includes an implementation of DIFFSTD, we will need to write it ourselves. We will start like any other detection approach, building a `run_tests()` function that takes a DataFrame in the `src\app\models\multi_timeseries.py` file.

Series Processing

Our users will send us a single list of all data points, keyed by some `series_key`. The first thing we'll need to do is to split out each series so that we can operate on them individually. Listing 16-2 shows how we can use a *list comprehension* in Python to do this.

Listing 16-2. A list comprehension that retrieves data for each series

```
series = [y for x, y in df.groupby("series_key", as_index=False)]
num_series = len(series)
l = len(series[0])
```

In this list comprehension, we group the data by series key and set `as_index=False` to ensure that we get the series key in the resulting list of DataFrames. From there, we see how many individual series we have as well as the length of the first series. This is a direct implication from the first assumption we follow, that each series has an equal number of data points matching exactly the same dates and times. We do not perform any length or date checking to ensure this and consequently have no built-in error correction.

Segmentation

Once we have the series split out, the next step is to break each series into segments of approximately seven data points. The reason we choose seven as our series length is that it guarantees we will have at least two segments per series, as each series must have at least 15 data points. As we split out data into segments, we will need to be cognizant of how many data points we include in each segment. This is necessary for two reasons. First, based on the formula for DIFFSTD, we want at least three data points in any given segment; if we end up with a segment containing just one or two data points, the formula will not make sense. Second, we want to ensure that each segment is fairly well balanced in terms of the number of data points. For example, we would not want a segment with 4 data points and another segment with 16. Although there is nothing in the DIFFSTD formula that requires segments be the same length, ensuring that segments are approximately the same size will make more intuitive sense to the end user.

Listing 16-3 shows the process we use to create our segments as well as checks on how many segments we have in total for each time series (segments).

Listing 16-3. Splitting each series into segments of size 6–10

```
series_segments = [np.array_split(series[x], (l // 7)) for x in range(num_
series)]
num_segments = len(segments[0])
```

On the first line, we use NumPy to split each series into an array of segments. We take the length of the series (the variable l) and perform integer division (//), dividing the total length of the series by 7. This tells NumPy the number of segments we want to create, and it allocates results accordingly. With that result, we have a stable set of segments for any length l >= 15, as well as a guarantee that we will never have fewer than six data points in a segment nor more than ten. We then calculate the number of segments within each series.

Comparing the Norm

After generating segments, we can run the DIFFSTD formula for each segment. The next challenge is, how do we turn a pairwise process like DIFFSTD into something that supports more than two series? We could perform every pairwise comparison and aggregate these together, but to simplify the problem, we will create a new series that represents the mean of all data points at each time. We can then compare each series against the newly generated mean series. This can work reasonably well because of the second assumption we make, that each series reflects the same underlying phenomenon and we expect things to move in approximate synchrony. Therefore, even if series operate on values of differing orders of magnitude, we can track changes in each series vs. the mean and see if elements move in approximately the same direction.

Listing 16-4 takes us through the process of generating segment means. We want to create an array of means, one per segment. The following code loops through each segment, figuring out the mean.

Listing 16-4. The function to generate means for each segment

```
def generate_segment_means(series_segments, num_series, num_segments):
  means = []
  for j in range(num_segments):
    C = [series_segments[i][j]['value'] for i in range(num_series)]
    means.append([sum(x)/num_series for x in zip(*C)])
  return means
```

Figure 16-1 shows how this function works in practice. In this case, the value for num_series is 3, and the value for num_segments is 4. For each data point in each segment, we will calculate the mean of all data points across all series at the same point in time. Thus, if the first data point of the first segment of the first series is 1, the second series is 4, and the third series is 7, we take the mean of these three values, which ends up being 4. Then we repeat the process for each data point in each segment, thereby guaranteeing that we have a means series that has the same size and shape as the individual series.

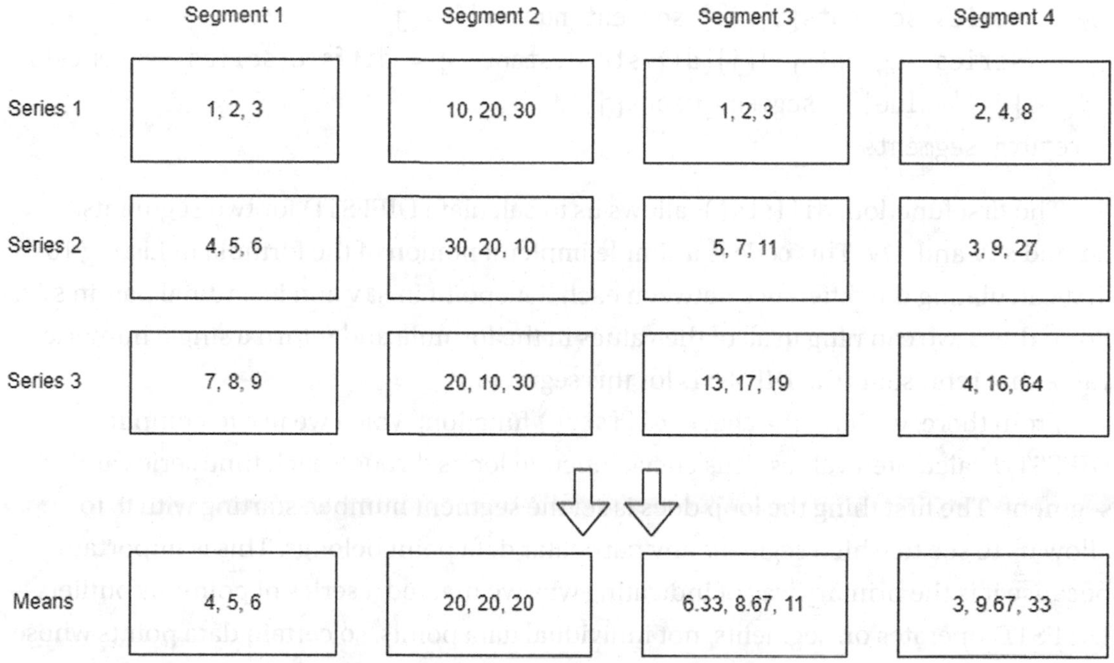

Figure 16-1. *Calculating the per-segment means across three series*

In the code, we call this resulting series segment_means. Once we generate this series, we can run DIFFSTD between the mean and each series. Listing 16-5 provides the code for the check_diffstd() function and its helper diffstd() function.

Listing 16-5. The code to run DIFFSTD against multiple time series

```
def diffstd(s1v, s2v):
  dt = [x1 - x2 for (x1, x2) in zip(s1v, s2v)]
  n = len(s1v)
```

```
  mu = np.mean(dt)
  diff2 = [(d-mu)**2 for d in dt]
  return (np.sum(diff2)/n)**0.5

def check_diffstd(series_segments, segment_means, num_series, num_
segments):
  for i in range(num_series):
    for j in range(num_segments):
      series_segments[i][j]['segment_number'] = j
      series_segments[i][j][diffstd_distance'] = diffstd(series_segments[i]
      [j]['value'], segment_means[j])
  return segments
```

The first function, `diffstd()`, allows us to calculate DIFFSTD for two segments, named `s1v` and `s2v`. This code is a simple implementation of the formula in Listing 16-1, first calculating the difference between each data point in `s1v` and its ordinal pair in `s2v`. From there, we can plug in all of the values in the formula and return a single numeric value that represents the DIFFSTD for this segment.

From there, we have the `check_diffstd()` function, which we use to compile DIFFSTD calculated values. This check function loops through each time series and segment. The first thing the loop does is set the segment number, starting with 0, to allow us to see to which segment any particular data point belongs. This is important because it is the primary way of indicating why we marked a series of points as outliers— DIFFSTD operates on segments, not individual data points, so certain data points whose behavior seems to be normal might be marked as outliers because they are part of a segment that diverges significantly, even if some of the data points within the segment might not diverge much from their peers.

The second thing the function does is calculate the DIFFSTD between this segment and its appropriate segment mean. We track that as a new column called `diffstd_distance`. This is akin to `anomaly_score,` but we want to reserve that term because we will extend the multi-series time series model in the next chapter and will use `diffstd_distance` as one component in the `anomaly_score` calculation.

Once we have calculated the segment numbers and scores for each segment, we need to return a single DataFrame. Right now, what we have is a list (series) of lists (segments) of DataFrames (individual data points). To combine them together, we will follow the two-step process in Listing 16-6. First, we want to flatten the list of lists, giving us a simple list of DataFrames. We can use a list comprehension to take care of this task, giving us back a list of segments containing one DataFrame per segment. From there, we use the concat() function in Pandas to combine together a list of DataFrames into one DataFrame and return it to the caller.

Listing 16-6. Turning a list of lists of DataFrames into a single DataFrame. We flatten the list of lists into a single list and then concatenate each DataFrame in the list together.

```
segment_means = generate_segment_means(series_segments, num_series, num_
segments)
segments_diffstd = check_diffstd(series_segments, segment_means, num_
series, num_segments)
flattened = [item for sublist in segments_diffstd for item in sublist]
df = pd.concat(flattened)
```

Once we have a single DataFrame containing the anomaly score per data point, it's time to figure out which of these points (by way of their segments) are outliers.

Determining Outliers

When it comes to determining whether a value of DIFFSTD is sufficiently different to indicate a break in series behavior, it is important to note that there are no hard-and-fast guidelines. The best we can do is compare the DIFFSTD for segments against other segments and create a relative ranking. This is similar to what we ran into with multivariate analysis techniques such as LOF and COF, though because the values of DIFFSTD will likely depend in part on the magnitudes of values in users' datasets, we will not be able to provide a "good enough" cutoff as a hard value.

What we can do instead is look for cases in which the DIFFSTD of a given segment is sufficiently far away from the mean value of all segments. Starting with a maximum distance score of 1.5 * mean from the mean, we can throttle this distance down using the sensitivity score, subtracting the sensitivity score (after scaling it to a value ranging from 0.00 to 1.00) from 1.5. For example, suppose the mean DIFFSTD value is 10.0. If we have a sensitivity score of 0.0, we would need to find a segment whose DIFFSTD is at least 1.5 * 10.0 away from the mean, or a minimum value of 25.0. At the other extreme, a sensitivity score of 100.0 would mean that any value above 10.0 + (1.5 – 1.0) * 10.0 or 15.0 would flag our outlier detector. Listing 16-7 shows how we can calculate the sensitivity threshold in the score_results() function.

Listing 16-7. Calculate the sensitivity threshold in Python using distance from the mean as our indicator. We do not use the `tests_run` dictionary, but it is there for future expansion, such as if we decide not to run DIFFSTD as part of our ensemble.

```python
def score_results(df, tests_run, sensitivity_score):
  series = [y for x, y in df.groupby("series_key", as_index=False)]
  num_series = len(series)
  diagnostics = { }
  for i in range(num_series):
    diffstd_mean = series[i]['diffstd_distance'].mean()
    diffstd_sensitivity_threshold = diffstd_mean + ((1.5 - (sensitivity_
    score / 100.0)) * diffstd_mean)
    series[i]['diffstd_score'] = (series[i]['diffstd_distance'] - diffstd_
    sensitivity_threshold) / diffstd_sensitivity_threshold
    series[i]['anomaly_score'] = series[i]['diffstd_score']
    # diagnostics segments removed
  return (pd.concat(series), diagnostics)
```

The first step is to split out each series, as we are dealing with a full DataFrame. Then, for each series, we calculate the mean of DIFFSTD values across all data points in that series. Once we are in possession of a mean, we can calculate the DIFFSTD sensitivity threshold using the preceding formula. One important note with our sensitivity threshold calculation is that 0.5–1.5 times the mean distance from the mean is not a hard-and-fast rule backed by academic study; it is simply a reasonable starting point. We could tweak this by calculating MAD from the median, calculating MAD from some quantile (like with box plots), or changing the magnitudes of distance from the mean.

Given a value for DIFFSTD sensitivity threshold, we now have enough information to generate `diffstd_score`, which is how far away a particular data point is from the threshold, normalized to the threshold. Our reporting is at the individual data point level, but we should keep in mind that we calculated the DIFFSTD distance at the segment level, so all data points within a segment have the same value for distance and will therefore have the same score. Because we are only using one algorithm (so far!) for multi-series time series detection, we will set the anomaly score equal to the DIFFSTD score and return all time series concatenated together.

Once we calculate the sensitivity threshold, we can track the max fraction of outliers using similar techniques to what we have used for the other input types, though we will want to track the max fraction of outliers per series, not as a whole. Listing 16-8 shows how we can do this in the determine_outliers() function.

Listing 16-8. Introducing the max fraction of outliers and making the final outlier determination

```
def determine_outliers(df, max_fraction_anomalies):
  series = [y for x, y in df.groupby("series_key", as_index=False)]
  max_fraction_anomaly_scores = [np.quantile(s['anomaly_score'], 1.0 - max_
  fraction_anomalies) for s in series]
  # diagnostics information elided
  sensitivity_thresholds = [max(0.01, mfa) for mfa in max_fraction_
  anomaly_scores]
  for i in range(len(series)):
    series[i]['is_anomaly'] = [score >= sensitivity_thresholds[i] for score
    in series[i]['anomaly_score']]
  return (pd.concat(series), diagnostics)
```

For each series, we want to find the percentile that corresponds to our max fraction of anomalies, so if max_fraction_anomalies is 0.1, we want the 1.0 – 0.1 = 90th percentile of anomaly scores. Then, we check to see if the max fraction of anomalies score is at least 0.01. If it is below 0.01, we use that cutoff instead because a sensitivity threshold of 0 or below indicates that this segment is not an outlier according to DIFFSTD.

Finally, for each data point in each series, we check if the score is greater than the sensitivity threshold for that series and, if so, set the value of is_anomaly to True; otherwise, we set its value to False. At this point, we have implemented time series analysis for multiple linked series. From here, we will want to put together sets of tests as well as an update to the Streamlit site.

Test and Website Updates

We have a new input type, and so we will need new tests and site updates. In this section, we will look at each of unit tests, integration tests, and website updates in turn.

Unit Tests

The unit tests for multi-series time series analysis look very similar to tests for single-series time series analysis, in that we will focus on ensuring that marginal differences in sensitivity score and max fraction of anomalies behave the way we expect them to: as sensitivity score or max fraction of anomalies goes up, the number of outliers marked should increase; as sensitivity score or max fraction of anomalies goes down, the number of outliers marked should decrease.

To perform these tests, we will use a sample input set containing data from two series. Each series contains 17 data points, ensuring that we have different numbers of data points per series. The first segment of each series is a direct linear translation: take the value from series s1 and add 10 to get the value for series s2. The second segment introduces variance in the process to ensure that we do have something that will show up in our tests. Listing 16-9 shows the code associated with the sensitivity score test case. We include a set of sensitivity scores ranging from 0 to 100 and expect either 0 or 16 results back, depending on our sensitivity score. We fix max fraction of anomalies to 1.0 to ensure that all outliers show up in our result set.

Listing 16-9. Testing marginal changes in sensitivity score while keeping max fraction of anomalies constant

```
@pytest.mark.parametrize("df_input, sensitivity_score, number_of_
anomalies", [
  (sample_input, 100, 16),
  (sample_input, 90, 16),
  (sample_input, 80, 16),
  (sample_input, 70, 16),
  (sample_input, 60, 16),
  (sample_input, 50, 0),
  (sample_input, 40, 0),
  (sample_input, 25, 0),
```

```
    (sample_input, 5, 0),
    (sample_input, 1, 0),
    (sample_input, 0, 0),
])
def test_detect_multi_timeseries_sample_sensitivity(df_input, sensitivity_
score, number_of_anomalies):
    # Arrange
    df = pd.DataFrame(df_input, columns=["key", "series_key", "dt", "value"])
    max_fraction_anomalies = 1.0
    # Act
    (df_out, weights, diagnostics) = detect_multi_timeseries(df, sensitivity_
    score, max_fraction_anomalies)
    print(df_out.sort_values(by=['dt']))
    # Assert
    assert(number_of_anomalies == df_out[df_out['is_anomaly'] == True].
    shape[0])
```

The reason we jump from 0 to 16 is twofold. First, the number of outliers is the number of *data points* marked as outliers. Because we focus on segments when working with multi-series time series analysis, we guarantee one of two cases: either all data points in the segment will be outliers or no data points in a segment will be outliers. Further, because we only have two series, they will necessarily mirror one another in terms of variance from the mean: the mean is the midpoint between these two series, so in order for one series to be sufficiently distant from the mean to be an outlier, the other must necessarily be so as well. Once we add in a third series, this statement will no longer hold true, but we are only working with two series in the unit tests. From this set of test results, we can understand that with a sensitivity score ranging from 0 to some value between 50 and 60, no segment's DIFFSTD will be sufficiently high as to become an outlier. Then, by the time we reach 60, each series will have one segment containing eight data points that is sufficiently different and therefore an outlier.

Integration Tests

For integration testing, we want to use a more realistic set of data, one which we expect to fit our three assumptions. As a brief reminder, the three assumptions are as follows: first, each time series is the same length with the same measured dates; second, each time series tracks the same underlying phenomenon; third, no time series has a dominant lagged correlation with any other series. The dataset we will use for our Chapter 16 Postman tests is Consumer Price Index (CPI) data, particularly the data in the Ch16 - Time Series - Multiple - No Debug - CPI Data test. We introduced the concept of CPI in Chapter 13, but we will recap it here. The US Bureau of Labor Statistics is responsible for surveying households and firms across the country on a monthly basis, asking a series of questions around the prices of certain products. They tabulate this data in an easy-to-analyze form called the Consumer Price Index, which tracks the changes in prices over time. Tracking prices is very useful, but it can be difficult to analyze prices of products differing in multiple orders of magnitude: a product whose price went from $1 per unit to $1.10 saw a 10% increase. If we try to plot that on the same graph as a new car whose price moved from $33,000 to $36,300, it will be difficult to see that both actually moved by the same percentage amount. This is one reason why the Bureau of Labor Statistics tracks CPI as percentage changes in goods. The other reason is that CPI is intended to act as an indirect gauge of inflation, which is itself a concept we work with in terms of percentage change over time.

To tie this all to percentages, the Bureau benchmarks a specific time frame as 100.0. This time frame will differ depending on the series—many series were benchmarked at 100.0 as the average of values between the years 1982 and 1984, for example. The specific benchmark does not itself matter for our purposes, as we can use the values of the index and see how they change over time as an indication of market trends (for a single product or small number of products) or inflation (across the board). In this case, we will use four CPI series as our test case. These four series represent nonseasonally adjusted prices of recreation services for urban consumers in the four regions of the country the Bureau of Labor Statistics uses: Northeast, South, Midwest, and West. Each of these series started in December of 1997 with a value of 100.0 and includes monthly data through July of 2021. Figure 16-2 shows each series over time. We can see that the four series move independently but tend to move in the same general direction. This is good for us because it lends credence to the idea that they are all tracking the same underlying phenomena: inflation and the interplay of supply and demand in recreation services.

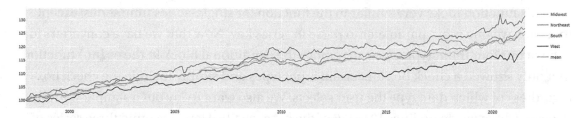

Figure 16-2. *Price changes in the cost of recreation goods and services over time, broken out by region. This is an example of the type of data we can use DIFFSTD to analyze*

The tests for this chapter use a sensitivity score of 60 and a max fraction of anomalies of 0.25. Otherwise, aside from the specifics of the JSON body we send and the endpoint we call, the tests are very similar to other API tests we have covered in the book.

Website Updates

In updating the Streamlit website, we will run into a couple of challenges. The first challenge is how we can display the results most effectively. With univariate data, we showed a bar chart of value vs. anomaly score. With multivariate data, we plotted the results as a bar chart of key vs. anomaly score. Single time series data included a combination of a gray line punctuated by scatter dots whose color and shape indicated to us whether we hit a change point. Because we have multiple series of data to show, we will need to take yet another approach to solving the problem of meaningful display of data. Before we do that, we should perform a little bit of clean-up work around converting lists to JSON. Listing 16-10 shows a new function that converts a multi-series time series Python list into JSON.

Listing 16-10. Convert an array of multi-series time series data into JSON for processing.

```
@st.cache
def convert_multi_time_series_list_to_json(time_series_str):
  mv_ast = ast.literal_eval(time_series_str)
  return json.dumps([{"key": k, "series_key":sk, "dt":dt, "value": v} for
  idx,[k,sk,dt,v] in enumerate(mv_ast)])
```

This function looks very similar to the function for single-series time series, except that we have one more parameter to parse in series key. Now that we have converters for each of our four input types, we can remove the conditional block in the main() function that only showed a check box for certain input types and simply include the check box regardless of which data type the user selects. We also want to lay out a sample starting dataset, which we do on lines 124–176 of the site.py file. Here, we build three series of data, specially crafted so that the second segment of series 1 and 2 may be outliers but not for series 3. This helps users see that we take outlier status by segment *by series*, not just by segment.

Once we have the data in place, we will want to display results. Listing 16-11 includes the code to display results. In this, we want to include each series as its own line with its own color. Because we will not know how many colors we need to use, we will use px.colors.qualitative.Safe to define our palette. These are colors that are safe for various forms of color vision deficiency, and as long as we do not have an extreme number of series to plot, there should be no visual overlap for most viewers.

Note   Not all colors in the Safe palette are monochrome-safe, meaning that you might have difficulty distinguishing colors in this grayscale book. To assist the reader, subsequent images that rely on understanding which series is which will include appropriate indicators.

Listing 16-11. Display the results of multi-series time series analysis, including the mean value at each time.

```
st.header('Anomaly score per segment')
df_mean = df.groupby("dt", as_index=False).mean("value")
df_mean['series_key'] = "mean"
ml = px.line(df_mean, x=df_mean["dt"],y=df_mean["value"],
        color=df_mean["series_key"], markers=False)
ml.update_traces(line=dict(color = 'rgba(20,20,20,0.45)'))
l = px.line(df, x=df["dt"], y=df["value"],
        color=df["series_key"], markers=False,
        color_discrete_sequence=px.colors.qualitative.Safe)
s = px.scatter(df,
```

```
    x=df["dt"],
    y=df["value"],
    color=df["series_key"],
    symbol=df["is_anomaly"],
    symbol_sequence=['square', 'circle'],
    hover_data=["key", "anomaly_score", "segment_number", "is_anomaly"],
    render_mode="svg",
    color_discrete_sequence=px.colors.qualitative.Safe)
s.update_traces(marker_size=9, showlegend=False)
g = go.Figure(data=l.data + s.data + ml.data)
st.plotly_chart(g, use_container_width=True)
```

The first thing we do in the code block is to calculate the mean at each date and time. This will allow us to plot a mean line in gray. We choose gray as the color because the mean line itself does not exist and should not be more eye-catching than the actual data, yet we do want some indicator of it so we can understand to what extent a particular series is diverging from the rest.

After tracking the mean, we create one line for each series, using the Safe palette in Plotly's qualitative palette list. Finally, we overlay the lines with a scatter plot indicating whether the data point is an outlier. As with single-series time series data, we will use squares to indicate that a data point is not an outlier and circles to indicate that it is an outlier. For more information, a user may hover over a specific point and see details on the key, anomaly score, segment number, and whether a data point is an outlier. Figure 16-3 shows a zoomed-in look at our CPI data from January of 2018 through July of 2021. In this figure, we can see four outlier segments, one for each region. In three of the outlier segments, we see the region rising faster than the others. The Midwest, however, saw a sharper decline in price during the middle of 2020 compared to other regions, although it did return to its prior trend line within a few months.

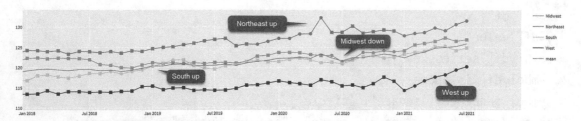

Figure 16-3. *Each series exhibits anomalous behavior between 2018 and 2020, made all the more interesting because there are only 11 total outlier-worthy segments over a 23-year period*

Conclusion

Over the course of this chapter, we began to put together the final piece of our outlier detection suite, introducing multi-series time series analysis. We created our own implementation of DIFFSTD, using it as we have used pre-built libraries in prior chapters. To support this part of the API, we also incorporated unit tests and integration tests. Finally, we updated the Streamlit website to allow users to try out multi-series time series analysis.

In the next chapter, we will introduce one more technique for multi-series time series analysis: Symbolic Aggregate Approximation. We will see how it differs from techniques like DIFFSTD and incorporate it into our multi-series engine.

## CHAPTER 17

Symbolic Aggregate Approximation (SAX)

The prior two chapters introduced a variety of calculations intended to compare two time series and gain an understanding of how far from alignment the two series are. We subsequently adapted this concept to handle multiple series by finding the mean behavior over fairly short segments of approximately seven data points. This is what we implemented as our multi-series analysis technique. In this chapter, we will take a technique that was originally designed to find recurring patterns within a single series and use it to extend our multi-series time series outlier detection engine. Along the way, we will once again create an ensemble of results and use that to inform our scoring of outliers.

What Is SAX?

Symbolic Aggregate Approximation (SAX) is a technique that Jessica Lin and Eamonn Keogh, along with other researchers then at the University of California – Riverside, developed. In their original 2002 paper, Lin et al. develop out the process of finding similar patterns of behavior within a time series even when the exact magnitudes of values differ due to factors like trend and seasonality. In other words, SAX potentially finds patterns in what classic time series decomposition would consider noise, as these patterns do not occur regularly and are not part of the trend or seasonal behavior.

Figure 17-1 shows an example of the type of pattern SAX can find. In this series, we can spot the same pattern five separate times, including three times in a row. Although this is a simple scenario in which we don't see any trend, seasonality, or noise, SAX is capable of finding a common pattern through these distractions.

© Kevin Feasel 2022
K. Feasel, *Finding Ghosts in Your Data*, https://doi.org/10.1007/978-1-4842-8870-2_17

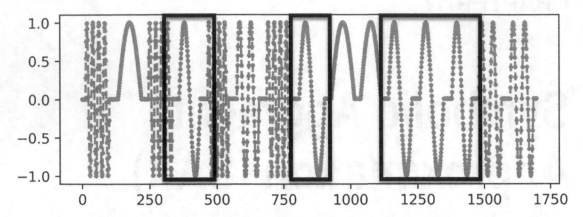

Figure 17-1. *An example of a pattern that occurs several times over the course of a time series dataset. Although this was an artificially generated dataset, tools like SAX allow us to perform this analysis on complex, real-world time series data*

Motifs and Discords

What we described in Figure 17-1, Lin et al. (2002) have called motifs. A *motif* in literary or musical terms is a recurring, dominant theme. The same concept applies here: a motif is a pattern that appears multiple times over the course of a time series. The pattern does not need to repeat itself at exactly the same magnitude, for exactly the same number of data points, or move in exactly the same ways, just so long as on the whole, the pattern is "close enough." In that way, the concept of a motif aligns very well with the Gestalt principles we learned about in Chapter 2, as we are attempting to identify patterns programmatically in a way that our eyes could do automatically. The difference is that humans get tired and have difficulty parsing nuance from large sets, whereas computers excel in this regard.

The opposite of a motif is a discord, something Keogh and Lin define in Keogh et al. (2005). A *discord* is a sequence that is sufficiently different from other sequences to merit explicit mention. Discord analysis works best with highly seasonal data but does not require exact periodicity. For example, suppose we track the response time for some given API endpoint. Overnight, when there are fewer callers and less system load, we might find the 95th percentile of response time to be 20ms. As more users sign in to begin the day and use our system, that response time might increase throughout the day to 200ms per call, drop during lunchtime, pick back up later in the afternoon, and tail off during the evening. Assume we have a lengthy time series of response time covering, for example, ten-minute periods over the course of several months. We would expect a rather consistent pattern, but sometimes we don't see that pattern. We might see the response time be 20ms throughout the entire day and later discover that on holidays, we see a tiny fraction of the total amount of work. We might see times spike well over 200ms per call on the day one of the load-balanced servers failed and we had to make do with just one server. The primary benefit of discords is that they give us an opportunity to ask, "What was special about that period of time? What caused this difference in behavior?" We can also easily apply the concept of discords to single-series outlier detection when we do expect reoccurring patterns.

Subsequences and Matches

The intuition behind SAX is fairly straightforward. Given a time series, we can define some *subsequence* length, where a subsequence is a length of the time series significantly smaller than the entire series length. These subsequences are not exactly like the segments we created in the prior chapter—the purpose of segments was to split each time series into manageable chunks and compare series behavior over the same chunk. Our segments also follow a tumbling window pattern: if our segment length is 8, the first eight data points form a single segment; then, the next eight form the next segment; and so on, until we have made our way through the entire dataset. By contrast, subsequences do not *need* to follow a tumbling window pattern.

They do, however, need to be some minimum distance apart for us to consider a match. The intuition behind this is laid out in the notion of a *trivial match*. Suppose we have some pattern ranging from points 80 to 150 in our time series, which we'll define as p. Suppose we have another subsequence q, which also runs from points 80 to 150, such that p = q. It is trivially true that yes, p will match q because they both cover exactly the same sequence of data points. Now let's make q range from 81 to 151. This is still an extremely close match, but it's also not providing us any useful information.

Therefore, a *nontrivial match* must be some minimum distance away from the original subsequence, so there can be no overlap between p and q. This means that in practice, subsequences are mutually exclusive even though their original definition did not require it (Lin et al., 2).

The normal use case for SAX is to find the *1-motif*, that is, the motif with the greatest number of similar subsequences and therefore the most commonly reoccurring pattern in a single time series. Incidentally, this implies that there can also be *2-motifs*, *3-motifs*, and so on. In our scenario, we will pay less attention to the notion of motifs within a time series and will instead use SAX to compare parts of multiple time series for commonality. To extend the musical analogy, we are focusing not on the common motifs of a performance but rather if the band is playing in concert.

Discretizing the Data

SAX starts with a technique known as *Piecewise Aggregate Approximation* (PAA). What this does is break a set of data points down into fixed-width frames, very similar to what we did with segments for DIFFSTD. We then calculate the mean of each frame and thereby reduce the number of data points we need to care about. Along with PAA, we also want to make sure to normalize the dataset and turn it into something following a Gaussian distribution. The reason we want to follow a Gaussian distribution is that this will allow us to create some arbitrary number of breakpoints in the distribution, with the idea being that each section contains an equal amount of area under the curve and therefore, an arbitrary data point is equally as likely to show up in any section. Figure 17-2 shows an example of this in action.

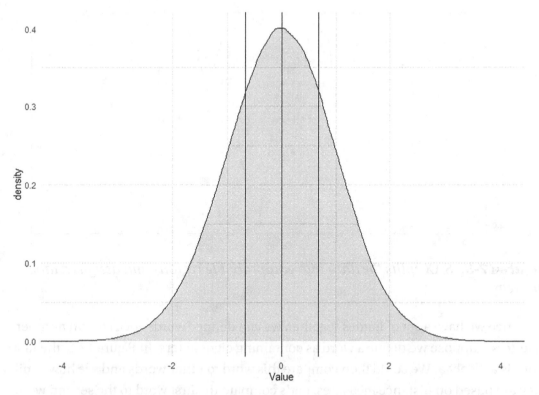

Figure 17-2. *This image splits a normal distribution into four regions, each of which has an equivalent share of the area under the curve. The splits are at -0.67, 0, and 0.67*

Lin et al. show the breakpoints for dividing a Gaussian distribution into anywhere from three to ten equiprobable regions, but subsequent research has indicated that three to five regions is generally all you will ever want and the "industry standard" for SAX is to use four regions, which is what we see in Figure 17-2.

Once you have a dataset following the normal distribution (i.e., mean of 0, standard deviation of 1), we can define each region as a letter and implement our "alphabet." In other words, with four regions, we need four letters, which we typically refer to as A, B, C, and D. These letters allow us to chain together combinations of PAA frames. For example, Figure 17-3 shows an example of this in action in which we average out, using PAA, a set of normalized data points. Then, we assign a letter to each data point based on the zone corresponding to each data point.

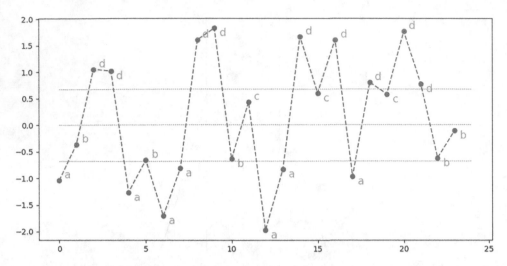

Figure 17-3. *SAX splits the data into equiprobable regions and assigns a label to each*

Once we have a set of frames together, we can define "words" of a certain number of letters. Suppose we define a word as containing eight letters. In Figure 17-3, the first word is abddabaa. We could then compare this word to other words and see how similar they are based on distance measures. Let's compare the first word to the second word, which is ddbcaadc. We can tell from a glance that the two are in fact different, but how do we calculate a proper distance? The answer comes from the values of the breakpoints between regions. Using an alphabet of four letters, we need to define three breakpoints, which represent the 25th, 50th, and 75th percentiles of the normal distribution's cumulative distribution function (CDF). Skipping over the math involved, the three breakpoints are at -0.67, 0, and 0.67. In other words, with our four-letter alphabet, we mark any point as A if its value is less than -0.67. We mark it as B if its value is anywhere from -0.67 to 0 (exclusive), C if it ranges from 0 to 0.67 (exclusive), and D if it is at least 0.67. By doing so, we will have approximately 25% of data points in each region.

We can also use these breakpoints to create a distance calculation. Table 17-1 shows the matrix we want to build in order to perform distance calculations.

Table 17-1. *Defining the distance between two data points. If we move no more than one letter away, the score will be 0*

	A	B	C	D
A	0	0	0.67	1.34
B	0	0	0	0.67
C	0.67	0	0	0
D	1.34	0.67	0	0

Moving more than one letter away has us increase by the breakpoint distance, which is 0.67 for each of our steps. This means that a move from A to D—or from D to A—constitutes a difference of 1.34. Lin et al. then define a minimum distance measure called MINDIST, whose formula we see in Listing 17-1.

Listing 17-1. The calculation of the minimum distance between two words

$$MINDIST\left(\widehat{Q}, \widehat{C}\right) = \sqrt{\frac{n}{w}} \times \sqrt{\sum_{i=1}^{w}\left(dist\left(\widehat{q}_i, \widehat{c}_i\right)\right)^2}$$

In Listing 17-1, n represents the number of data points per word. For the sake of providing an example, we will say that each letter represents the average of four data points, which we calculate via PAA. We have eight letters per word, leading us to 32 total data points per word. Next, w represents the length of our word, which is 8. For each pairwise comparison, we calculate the distance measure from our table, square the result, and take the square root of the summation. Then, we multiply that by the square root of the quantity n divided by w, providing us the final definition of MINDIST.

In comparing the two words, we have two instances in which letters are three apart (i.e., a–d or d–a), three instances in which letters are two apart, and three instances in which letters are no more than one apart. Summing the squares of distances, we end up with a value of $2 * 1.34^2 + 3 * 0.67^2 = 4.9379$. Taking the square root of this result, we get a value of 2.222. Meanwhile, 32 / 8 is 4 and the square root of 4 is 2, so if we multiply 2.222 by 2, we end up with a MINDIST measure of 4.444. MINDIST values are comparable across words with equal counts of data points and letters per word, and a higher value of MINDIST indicates that two words are "more different" than a lower value of MINDIST.

Now that we have the idea behind SAX, let's see what it takes to incorporate it into multi-series time series analysis.

Implementing SAX

There are several Python packages that offer SAX, including `tslearn`, `saxpy`, and `pyts`. In this section, we will implement SAX using the `tslearn` package, as it provides us some functionality that makes it easier to extend SAX to multi-series time series analysis, including a built-in distance calculation.

The first thing we will need to do is update the `run_tests()` function for multi-series time series analysis to incorporate SAX. In the prior chapter, we broke our incoming DataFrame out into each individual time series using a list comprehension. Then, we broke each series into segments of approximately seven data points apiece and called DIFFSTD on each segment, comparing one segment to the mean of the corresponding series' segments. We will not use the same segmentation process for SAX, however, because we have a different process available to us.

Segmentation and Blocking

In the `tslearn` package, the `SymbolicAggregateApproximation()` class takes three parameters: the number of segments (`n_segments`), the alphabet size (`alphabet_size_avg`), and whether to normalize the data before processing (`scale`). For the latter two parameters, we will hard-code the alphabet size to 4 and force normalization, as these are strong recommendations for us to use based on the existing SAX literature. The main parameter we need to consider is the number of segments. One important thing to note is that we cannot pass in our own mechanism for segmenting the data, so we should not assume that data points in one segment for DIFFSTD will necessarily be in the same segment for SAX. This turns out to be a good thing on the whole, as it allows us to fine-tune the outlier detection process. Suppose data points 1–7 and 8–15 constitute two segments for DIFFSTD. Meanwhile, SAX uses 1–5, 6–10, and 11–15. Further, for some series, suppose we end up with a very high score for DIFFSTD for points 1–7 and a low one for 8–15, indicating that the first segment is an outlier. Meanwhile, running SAX, we see no indication of an outlier for points 1–5, an outlier for 6–10, and no outlier for 11–15. We could draw a conclusion from this that points 6 and 7 are the most likely outlier points. For this reason, it can be beneficial for our different methods to use differently sized and shaped segments.

Note   This same logic also implies that we could potentially gain from performing DIFFSTD using differently sized segments and combining the scores together, similar to what we did with nearest neighbor determination and COF. Extending DIFFSTD to become its own ensemble may be an interesting exercise for the reader.

In determining our segment size for SAX, we should note that each segment will output a single letter, that we want some number of letters to constitute a word, and that we want enough words to allow us to make a sufficient number of comparisons between series. For this reason, I chose segment sizes based on the length of each time series, increasing with orders of magnitude. For cases with fewer than 100 data points, we make each segment contain two data points, the minimum to calculate a nontrivial piecewise average. If we have at least 100 but fewer than 1000, we will include three data points per segment. Finally, for 1000 or more data points, each segment will constitute five data points.

Listing 17-2 shows how to perform these calculations in the check_sax() function, which requires three parameters: series, representing our incoming time series; num_series, which tells us how many unique time series we have; and l, which represents the number of data points in each time series. We also need to convert each series into a list of data points containing only the value so that we match the required shape for inputs.

Listing 17-2. Calculating segment splits and shaping data in the necessary manner for SAX

```
def check_sax(series, num_series, l):
  if (l < 100):
    segment_split = 2
  elif (l < 1000):
    segment_split = 3
  else:
    segment_split = 5
  sax = SymbolicAggregateApproximation(n_segments= l//segment_split,
  alphabet_size_avg=4, scale=True)
  slist = [series[i]['value'].tolist()
    for i in range(num_series)]
```

Once we have our dataset prepared, we need to determine how large our words will be. There is no one-size-fits-all answer for this, but I chose 4 as the word size for three fairly simple reasons. First, we want a word size that is large enough to allow the likelihood of reasonable change. Second, we want a word size that is not so long that it obscures change. Finally, we know that the minimum number of data points in a series is 15, so a word size of 4 and a segment size of 2 mean we will always have at least two words to compare. This means each word will constitute 8 (if we have fewer than 100 total data points), 12 (100–999 total data points), or 20 data points. Listing 17-3 continues the check_sax() function and performs transformation and fitting into a sax_data array. The resulting array is actually an array containing a list of lists. Breaking this down, we have an array of elements. Each element is a list, one per series. Inside that list, we have each letter in its own list, with tslearn representing the four letters as the numbers 0, 1, 2, and 3 rather than the letters A, B, C, and D.

Listing 17-3. The process of fitting, transforming, and preparing to convert results into words for comparison. An example of the result appears as a comment above the fit and transformation function.

```
# array([ [[1],[1],[1]], [[1],[0],[2]], [[2],[2],[1]] ])
sax_data = sax.fit_transform(slist)
word_size = 4
num_words = len(sax_data[0])//word_size
```

In the case of the sample array in Listing 17-3, we have three letters in each series, though tslearn shows the letters as 0–3 rather than A–D. From there, we want to combine things into words of size 4, which means we will need to know how many of these words to create. To get this number, we perform integer division, dividing the total number of letters by the word size and returning the number of words we will need to create.

Note   In the preceding trivial example, we would have an issue because there aren't enough data points to constitute a full word. This is not actually a problem we will run into in practice, however, because we have a required minimum number of data points that guarantees at least four letters.

Making SAX Multi-series

So far, the way we have described SAX has been in the context of a single time series. Its concepts of motifs and discords indicate patterns (and breaks from the pattern) in a single series. To make this work for our multi-series problem, we will need to do a bit of work. We have at our disposal two key things: first, we have series of the same size and that are supposed to represent the same underlying phenomenon. Second, tslearn provides a distance_sax() function that calculates the pairwise distance between two words using the MINDIST formula from Listing 17-1. Using these two facts, we can break up each series into a set of words and make pairwise comparisons between each word and its corresponding matches in the other series. With the resulting distance measures, we can calculate the average distance from a word's peers and store it in a matrix, like we see in Listing 17-4.

Listing 17-4. Calculating the pairwise comparisons for each word and applying the results to data points in each series

```
m = np.empty((num_series, num_words))
for i in range(num_series):
  for j in range(num_words):
    m[i][j] = sum(sax.distance_sax(
        sax_data[i][j*word_size:(1+j)*word_size],
        sax_data[k][j*word_size:(1+j)*word_size])
        for k in range(num_series))/(num_series-1)

for i in range(num_series):
  series[i]['sax_distance'] = [m[i][min(j//(word_size*segment_split),
  num_words-1)]
for j,val in enumerate(series[0].index)]
```

The first thing to do is create an array with one row for each series and one column for each word. To populate that array, we calculate the SAX distance using a comprehension: for each series k, we compare the word ranging from j * word_ size up to `(1+j) * word_size` to the word in series i. Word size starts at 0, so our first comparison will run from `sax_data[i][0:4]` vs. each `sax_data[k][0:4]`. This will compare elements 0, 1, 2, and 3 to each other and then divide by the number of comparison series. The next run will compare `sax_data[i][4:8]` to each `sax_data[k][4:8]`, comparing elements 4, 5, 6, and 7 and then dividing by the number of comparison series. We do this until we run out of words to compare. One note is that if you compare a series to itself, the distance will always be 0, so we do not have to worry about removing that comparison from the numerator; we simply subtract 1 from the denominator to indicate that we do not wish to include the series itself in this comparison.

Once we have all of the elements of the matrix in place, the final command loops through each series and adds the `sax_distance` column to the series DataFrame, setting the value equal to the relevant word in the matrix. In the likely event that we have a few leftover data points that don't cleanly belong to a word—for example, if we have 19 data points for a segment size of 2 and a word size of 4—we set those data points' values to the final average distance in the series, which is why we need the `min()` function in the calculation. Those data points were used to generate the distances for the final word, so we want to make sure they get the same credit.

Now we have a SAX distance for each data point. In the next step, we will revise our outlier scoring mechanism to incorporate SAX distance as well as DIFFSTD distance.

Scoring Outliers

Now that we have another algorithm in our multi-series time series ensemble, we will need to enhance our two outlier determination functions: `score_results()` and `determine_outliers()`. Listing 17-5 shows the new `score_results()` function, which takes in a DataFrame and sensitivity score.

Listing 17-5. An updated function to generate a score based on DIFFSTD and SAX distances. Diagnostic code has been removed to simplify the code listing.

```
def score_results(df, tests_run, sensitivity_score):
  series = [y for x, y in df.groupby("series_key", as_index=False)]
  num_series = len(series)

  for i in range(num_series):
    diffstd_mean = series[i]['diffstd_distance'].mean()
```

```
diffstd_sensitivity_threshold = diffstd_mean +
    ((1.5 - (sensitivity_score / 100.0)) * diffstd_mean)
series[i]['diffstd_score'] = (series[i]['diffstd_distance'] -
diffstd_sensitivity_threshold) / diffstd_sensitivity_threshold
sax_sensitivity_threshold =
    max(100.0 - sensitivity_score, 25.0)
series[i]['sax_score'] = ((series[i]['sax_distance'] * 15.0)-
    sax_sensitivity_threshold) / sax_sensitivity_threshold
series[i]['anomaly_score'] = series[i]['sax_score'] +
    series[i]['diffstd_score']
return (pd.concat(series), diagnostics)
```

The function starts by breaking data out by series and then operating on each series independently. The calculation for DIFFSTD score is the same as in the prior chapter: we measure distance from a value 1.5 times away from the mean and will have a positive DIFFSTD score if we exceed that value.

The next two lines cover SAX. Just like DIFFSTD, SAX does not have a hard cutoff point, meaning that we do not have any solid guidance on what distance is sufficient to indicate that the word containing this data point is an outlier. In practice, though, we can see some amount of divergence when the word distance is approximately 2.5, and that divergence is significant at a size of 3 or 4. Putting this in the scale of our sensitivity score, which ranges from 0 to 100, we could multiply the SAX distance by 15, such that a distance of 2.5 becomes a value of 37.5. We then compare this to the value 100.0 - sensitivity score, so if the sensitivity score is 60, our threshold would be 40. Subtracting the augmented distance of 37.5 from our threshold of 40 and then dividing by the sensitivity threshold, we end up with -2.5 / 40, or a SAX score of -0.0625. Increasing the sensitivity score serves to decrease the threshold, and so a sensitivity score of 75 would turn our SAX score from -0.0625 to (37.5 - 25) / 25 = 0.5, a positive score indicating an outlier. Note that we cap the minimum sensitivity threshold at 25 to prevent absurd results, like SAX scores in the millions. Therefore, a sensitivity score above 75 would still result in us returning a SAX score of 0.5 in this case.

Finally, we sum the SAX and DIFFSTD scores and return that result for outlier determination. Listing 17-6 shows how we make this determination.

Listing 17-6. Determining which data points are outliers based on score and max fraction of anomalies. Diagnostic code has been removed to simplify the code listing.

```
def determine_outliers(
  df,
  max_fraction_anomalies
):
  series = [y for x, y in
    df.groupby("series_key", as_index=False)]
  max_fraction_anomaly_scores = [np.quantile(
    s['anomaly_score'], 1.0 - max_fraction_anomalies)
    for s in series]
  sensitivity_thresholds = [max(0.01, mfa)
    for mfa in max_fraction_anomaly_scores]
  for i in range(len(series)):
    series[i]['is_anomaly'] = [score >= sensitivity_thresholds[i]
      for score in series[i]['anomaly_score']]

  return (pd.concat(series), diagnostics)
```

Because we already deal with the notion of sensitivity score, this function does not need the sensitivity score to make a determination; it only needs our augmented DataFrame and the max fraction of anomalies setting. The first step of the function is, like in the prior chapter, to find the cutoff point for max fraction of anomalies. As an example, if the user sends a value of 0.15, we are looking for the 85th percentile of scores, ensuring that the outlier detection engine will mark no more than 15% of our results as outliers. Then, we calculate the sensitivity threshold. We put a floor for sensitivity threshold at 0.01, as we consider any score at or below 0 not to be an outlier. Negative scores are only possible when at least one of SAX or DIFFSTD determines that the point's distance is less than the threshold we set using the sensitivity score, so we can focus on the points that are at least 0.01. We end by marking as outliers all of the data points whose scores meet or exceed the sensitivity threshold criterion.

With all of the calculations in place, we can update our Streamlit site and ensure that tests still work as we expect them to.

Test and Website Updates

In this section, we will briefly review the unit and integration tests and then cover a few minor website updates.

Unit and Integration Tests

By introducing SAX into the mix, we can expect the number of outliers to change, with SAX sometimes acting as a dampener and sometimes enhancing a marginal outlier signal. In our unit test scenario, SAX acts as a mild dampener. Using DIFFSTD alone, the function `test_detect_multi_timeseries_sample_sensitivity()` returns 16 outliers with a sensitivity score of 60. Introducing SAX pushes the threshold up a little bit so that 60 is not quite sensitive enough to mark these data points as outliers. All other tests in the sensitivity test case, as well as all tests in the max fraction of anomalies test case, remain the same.

As far as integration tests go, we do see a larger number of outliers here than in chapter 16 when reviewing the CPI dataset. In Chapter 16, we saw 77 data points marked as outliers, but now, we have 134. In this case, SAX picked up smaller shifts in behavior which DIFFSTD was not quite sensitive enough to catch. The large majority of these outliers take place in the West region, which helps explain why it remains consistently lower than other regions throughout the dataset. Figure 17-4 shows an example of this in the years 2016 and 2017, where we mark the West as an outlier every month for the better part of two years.

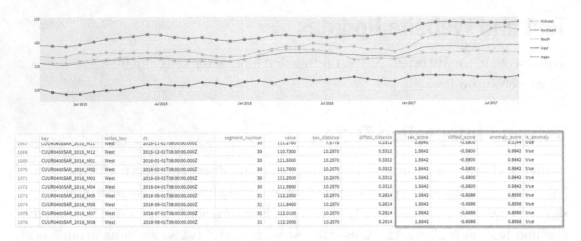

Figure 17-4. *From 2015 through 2017, changes in the cost of recreational activities in the western United States did not tend to follow the same pattern as the rest of the country*

Website Updates

Just as with tests, we require relatively few changes to the website in this chapter. The biggest change involves adding features like SAX and DIFFSTD distances and scores to the results table.

Conclusion

In this chapter, we adapted the concept of Symbolic Aggregate Approximation (SAX) to work for multi-series time series analysis. Although SAX is typically intended for single-series analysis, we created a process to perform word comparisons—that is, comparisons of segments of data points—across each series. This is because of one of the three core assumptions we made in the prior chapter: all of the time series in a single set should represent the same underlying phenomenon.

With the addition of SAX, our outlier detection engine is now complete! We have at least one method to analyze four of the most common use cases for outlier detection: univariate scenarios, multivariate scenarios, single-series time series analysis, and multi-series time series analysis. We can certainly tweak the code to make it faster and tighten up some of the logic behind some handwavy and arbitrary cutoff points we've used throughout the book for the sake of simplicity. That said, these alterations are more on the order of tweaks rather than wholesale changes to the underlying architecture.

Now that we have a product, it makes sense to test it against the competition, particularly if we have datasets with known outlier information. This will give us a better understanding of how we've done and where we can still improve. As we wrap up Part IV of this book and move into Part V, we will see how our detection engine stacks up to readily available commercial products.

PART V

Stacking Up to the Competition

In this final part of the book, we want to see how our anomaly detection engine stacks up to the competition. Chapter 18 will briefly review three competitors from public cloud providers Amazon, Microsoft, and Google. We will also set up an account with Microsoft's Azure Cognitive Services Anomaly Detector service and try it out. Finally, in Chapter 19, we wrap up the book with a bake-off, comparing our engine to the Cognitive Services anomaly detection service. We will review the results of this bake-off and close with ideas on how to make our anomaly detection engine and service better for users.

## CHAPTER 18

Configuring Azure Cognitive Services Anomaly Detector

Over the past dozen (or so) chapters, we have created a fairly complete outlier detection engine. With this in hand, it makes sense to compare our creation to some of the best general-purpose outlier detection engines available on the market today. In this chapter, we will review a few of the many options available for general-purpose outlier detection, focusing on cloud platform providers. Then, we will configure one of these for our own use: Microsoft's Azure Cognitive Services Anomaly Detection service.

Gathering Market Intelligence

There are a variety of outlier detection engines available to us. In this section, we will quickly review options available for three of the largest public cloud providers. One important factor in this section is that we will only cover explicit outlier detection capabilities in services. For example, all of the major cloud hosting providers we will review include general-purpose machine learning capabilities, so you could write and deploy your own outlier detection service—including the one we have built throughout this book—on any of the clouds we will look at here. Instead, we will focus on the capabilities of their pre-built outlier detection offerings.

Amazon Web Services: SageMaker

Amazon Web Services (AWS), as of the time of writing, remains the largest public cloud provider in the world. Therefore, it makes sense to look at its anomaly detection

313

K. Feasel, *Finding Ghosts in Your Data*, https://doi.org/10.1007/978-1-4842-8870-2_18

offerings first. These offerings are wrapped into its SageMaker service, which provides data scientists and developers a fully managed machine learning service. SageMaker brings a variety of algorithms to data scientists, allowing them to train and score models efficiently on AWS compute and then deploy these models for other services and applications to use.

The primary algorithm available for outlier detection in SageMaker is called *Random Cut Forest* (RCF), an algorithm somewhat similar to the isolation forest concept we touched upon in Chapter 9. As a brief refresher, tree-based approaches like isolation trees are good at indicating whether something appears to be an outlier but will not necessarily be able to provide us a directly comparable score. RCF in SageMaker mitigates this issue by providing us an anomaly score we can use for comparison in addition to a label indicating whether something is an outlier.

The primary use case for RCF in SageMaker is to perform single-series time series outlier detection, particularly streaming data coming from Amazon Kinesis streams. That said, it can also handle non-time series analysis as well. At the time of writing, it does not cover the multi-series time series scenario.

Microsoft Azure: Cognitive Services

The second-largest public cloud in the world is Microsoft Azure. Azure has several machine learning services available for public use, but the one that is most interesting to us is the Anomaly Detector service in Azure Cognitive Services. Azure Cognitive Services is a set of APIs offering pre-trained artificial intelligence services in a variety of areas, including speech, language, vision, and decision-making, the latter of which includes their Anomaly Detector service.

Microsoft's Anomaly Detector service focuses on time series data, meaning that it does not handle any sort of non-series data, whether univariate or multivariate. Within the realm of time series analysis, there are two options available to us. The first of these is univariate time series, which we have labeled as single-series time series. In conjunction with univariate time series analysis, Cognitive Services also offers multivariate time series analysis, which is the conjunction of multiple series of data. Importantly, this is not quite the same as the multi-series time series analysis we have worked with. One of the key assumptions of our multi-series time series analysis is that each series describes the same underlying phenomenon. By contrast, multivariate anomaly detection in Cognitive Services assumes that the multiple series interact in a common fashion.

Using an example to clarify the difference, we looked at Consumer Price Index movement for recreational services across different regions in the United States. This helped us get a feeling for common underlying phenomena: inflation, changes in supply of services, and changes in demand for services. Each series is independent, but we expect each series to move in a similar fashion to every other series and consider something to be an outlier when it moves discordant to the other series.

Multivariate anomaly detection in Cognitive Services, meanwhile, expects not that each series will move in the same fashion but rather that the series will move in a *correlated* fashion. Instead of looking at CPI across regions, a more apt use case for multivariate anomaly detection would be a set of sensors providing information on a piece of factory equipment, looking at various factors like core temperature, oil pressure, oil viscosity, oil temperature, and revolutions per minute to determine the health of that equipment. In a normal scenario, some measures will move in the same direction, but others may move in opposite directions: measures of oil pressure and oil viscosity are higher when the oil temperature is lower, so as our factory equipment's engine heats up, those numbers decrease. What Cognitive Services tracks in this case is whether combinations of values are outside the norm, not whether one series is outside the norm of all series.

In addition to Azure Cognitive Services, Azure Stream Analytics includes with it the capacity to perform anomaly detection on time series data using two techniques. We are already familiar with one of the two techniques: change point detection. The other technique looks at spikes and dips in time series data. These two techniques are also available outside of Azure Stream Analytics in the ML.NET library. These two latter services are outside the scope of this book, and so we will not cover them in any further detail, focusing instead on Cognitive Services.

Google Cloud: AI Services

A third significant cloud player is Google Cloud. Although not nearly as large as AWS or even Azure, it still takes a respectable share of the public cloud infrastructure market. Google's anomaly detection offerings tend to be features of other services, such as Google Analytics or Apigee, Google's solution for API management. Even so, Google's BigQuery ML includes the ability to detect time series and non-series anomalies. For non-series data, BigQuery ML has two options. The first option is a K-means clustering model, which identifies outliers as being sufficiently far from any cluster centroid.

The other approach involves an autoencoder, which builds a model and compares the distance between the predicted value and actual value. If that distance is sufficiently large, the point becomes an outlier.

For time series data, BigQuery ML supports a variant on ARIMA that they call ARIMA_PLUS. This time series model uses capabilities in ARIMA to decompose seasonality and trend as well as handle holidays in time series data. Similar to non-series datasets, this technique builds a model and describes an outlier as something sufficiently distant from the model's expected result.

Now that we have looked at three of the major public cloud players, let's focus in on one of these solutions: Azure Cognitive Services and its Anomaly Detector service.

Configuring Azure Cognitive Services

In this section, we will walk through the process of configuring Azure Cognitive Services and using its Anomaly Detector service. By the end of this chapter, we will have all of the pieces in place to compare our anomaly detection engine to the one in Cognitive Services.

Set Up an Account

Azure Cognitive Services does require a Microsoft Azure subscription. If you do not already have an Azure subscription, you can sign up for a free one-month trial at `https://azure.microsoft.com/free/` and get a one-month allocation of $200 USD in credits. Note that a free subscription does require a valid email address and credit card number. After your trial expires, you can retain free access to certain Azure services, including the Azure Cognitive Services Anomaly Detector. The free F0 pricing tier provides you access to univariate time series anomaly detection and allows you to make 20,000 transactions per month. The other pricing tier available is S0, which lets you make an unlimited number of univariate time series anomaly detection calls at $0.314 USD per 1000 transactions (as of the time of writing—Azure pricing is liable to change over time). The S0 tier also provides you access to the public preview of multivariate time series anomaly detection. Access to Azure services in public preview is typically free, and multivariate time series anomaly detection is no exception. Once the feature does reach general availability, Microsoft will announce pricing details, but as of the time of writing, the exact details on cost are not publicly available.

Once you have an Azure subscription, navigate to the Azure portal at `https://portal.azure.com`. From there, type "anomaly detector" into the Azure search bar and choose the *Anomaly detectors* entry from the resulting list, as shown in Figure 18-1.

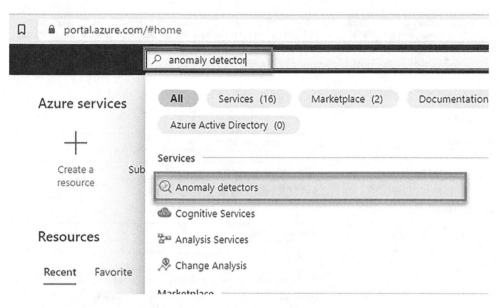

Figure 18-1. *Choose "Anomaly detectors" from the Azure services list*

Next, select the + *Create* option to create a new service. Choose the Azure subscription and resource group you would like to use to host your instance of the Cognitive Services Anomaly Detector service. You will also want to pick a region and name your service. Figure 18-2 shows an example of the service creation screen, including a breakdown of the two pricing tiers and an example of the error message you receive if you try to use a name somebody else has chosen. All Anomaly Detector service names must be globally unique, so you will need to enter a unique name.

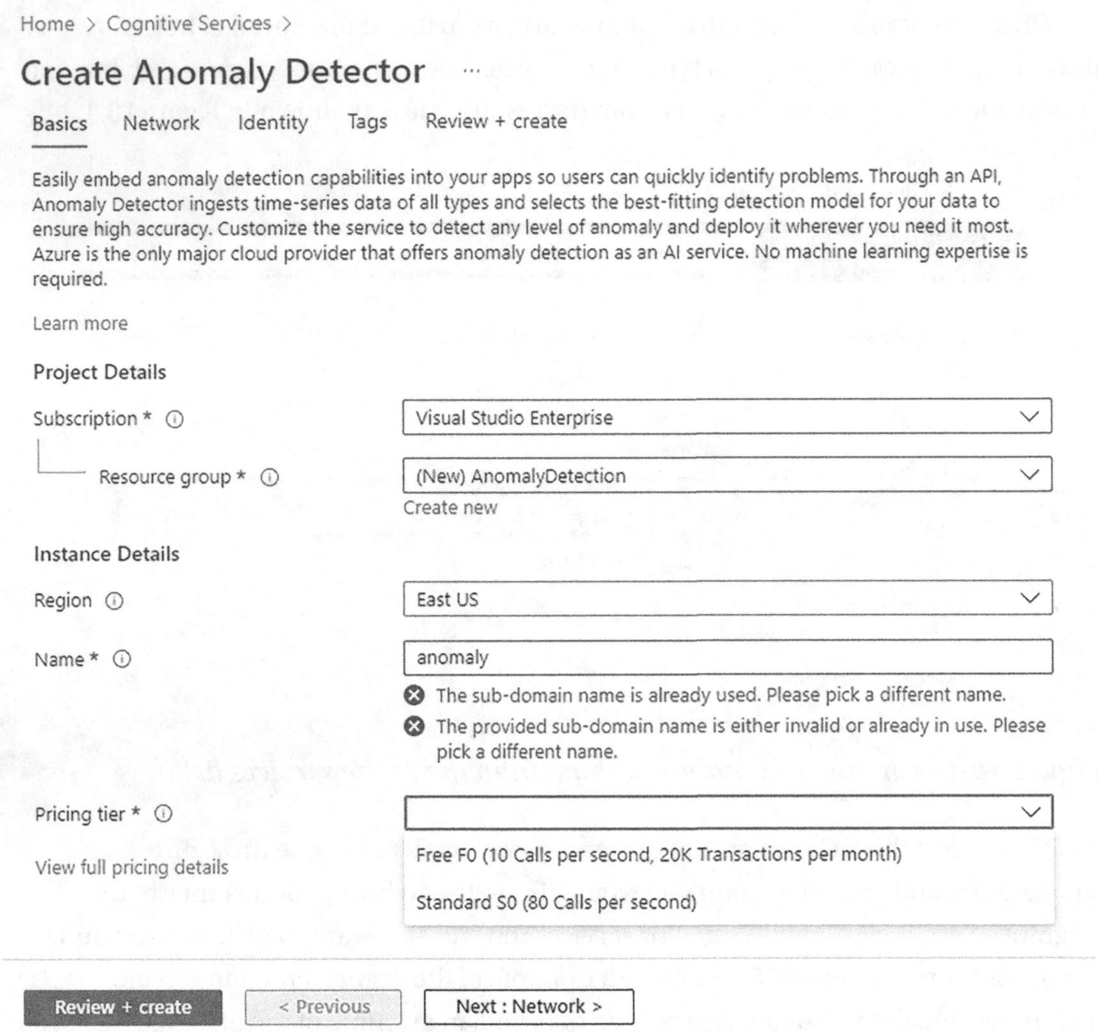

Home > Cognitive Services >

Create Anomaly Detector   ...

Basics   Network   Identity   Tags   Review + create

Easily embed anomaly detection capabilities into your apps so users can quickly identify problems. Through an API, Anomaly Detector ingests time-series data of all types and selects the best-fitting detection model for your data to ensure high accuracy. Customize the service to detect any level of anomaly and deploy it wherever you need it most. Azure is the only major cloud provider that offers anomaly detection as an AI service. No machine learning expertise is required.

Learn more

Project Details

Subscription * ⓘ

| Visual Studio Enterprise | ⌄ |

    Resource group * ⓘ

| (New) AnomalyDetection | ⌄ |

Create new

Instance Details

Region ⓘ

| East US | ⌄ |

Name * ⓘ

| anomaly |

  ❌ The sub-domain name is already used. Please pick a different name.
  ❌ The provided sub-domain name is either invalid or already in use. Please
     pick a different name.

Pricing tier * ⓘ

| | ⌄ |

View full pricing details

Free F0 (10 Calls per second, 20K Transactions per month)

Standard S0 (80 Calls per second)

[Review + create]   [< Previous]   [Next : Network >]

Figure 18-2.   Creating an Anomaly Detector service. The name must be globally unique, so choose something other than "anomaly"

Once you have filled out the first page of the form, you can select *Review + create* to finish creating the service. The other tabs on the creation form allow you to control infrastructure details for the service, such as limiting access to only specified Azure virtual networks or providing the service access to certain other Azure resources. For the purpose of comparing our anomaly detection service to this one, we will leave the networking section alone, meaning that the service is not limited to specific IP addresses or just machines on Azure's networks. Even so, your service will not be open to the broader world to abuse, as we will see shortly.

After creating the service, select it from the Cognitive Services grid. If you have trouble finding your service, repeat the search for "anomaly detector" in the Azure search bar and you should be able to find it. On the *Overview* tab for your service, you should be able to see the Endpoint, which is the URL to access your anomaly detector service. Copy this URL and save it somewhere convenient, as you will need it shortly. Figure 18-3 shows where you can find this endpoint information.

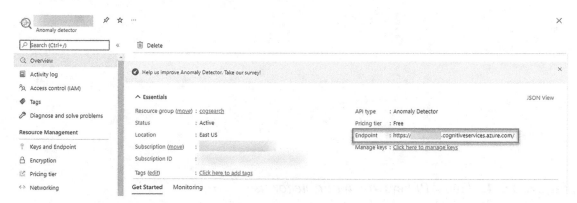

Figure 18-3. *Copy the endpoint URL that you see on the Overview tab*

In addition to an endpoint URL, we need an API key. You can find your API keys in the *Keys and Endpoint* tab, as in Figure 18-4. To prevent unauthorized service use (and expenditures), an API key is required to access the anomaly detector service. Cognitive Services generates two keys for you, allowing you to rotate between keys without service disruption. For now, copy one of the two keys to the same place you saved your endpoint URL.

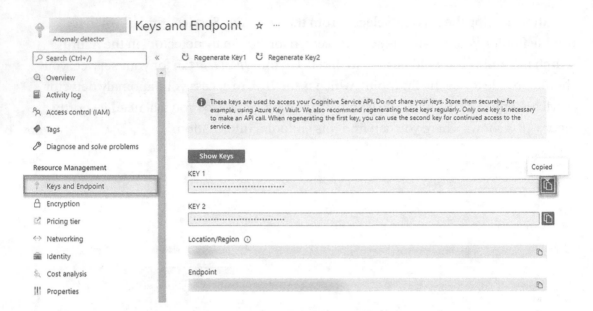

Figure 18-4. *Two API keys are available for use*

With an endpoint and an API key, we have enough detail to try out Microsoft's Anomaly Detection demo application.

Using the Demo Application

Microsoft provides access to an Anomaly Detector demo application to evaluate their service. You can access this website at `https://algoevaluation.azurewebsites.net`. This service does require you to enter your endpoint and API key in input boxes on the right-hand side. Then, you can choose a local file or try one of their pre-registered samples. Figure 18-5 shows an example of this service in action. At the top of the screen, we see three options for API usage: Last API, Entire API, and Multivariate API. The Last API option performs outlier detection over windows of data, and Entire API looks at the entire stream of data in one go. In this case, we are looking at the Entire API tab and Sample 2. Azure Cognitive Services includes two input parameters: sensitivity and max anomaly ratio, which are analogous to our sensitivity score and max fraction of anomalies.

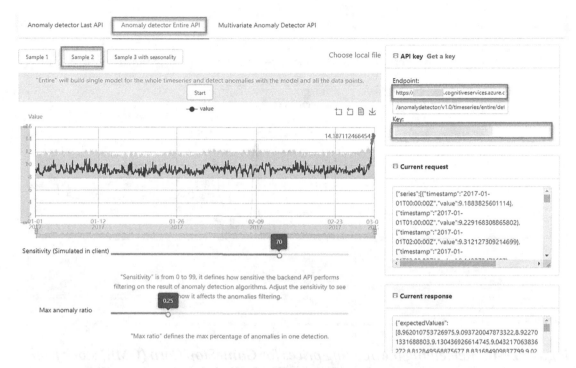

Figure 18-5. *An example of univariate time series anomaly detection using Azure Cognitive Services*

In addition to the sample datasets that are available, we can also upload small files and perform outlier detection on the data. In the /data/ folder of the accompanying code repository for this book, there is a GME.csv file that includes some of the stock price data for GameStop Corp we used in Chapter 14. The Azure Cognitive Services Anomaly Detector service can ingest this CSV data, read the date and open price (the first feature after the date in the file), and perform outlier detection on that set. Looking at Figure 18-6, we can see that Cognitive Services uses a model-based approach to outlier detection, marking data points as outliers when they go outside the predicted boundaries.

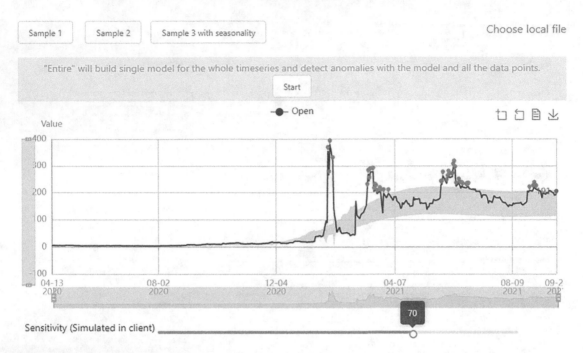

Figure 18-6. *Reviewing the opening prices for GameStop Corp (GME) stock from April of 2020 through September of 2021. The shaded area in the graph represents model expectations for closing price. The increased volatility of GME stock prices is apparent in the model's expectations*

This is in contrast to our anomaly detection engine, which focuses on change points rather than individual values outside the norm. Figure 18-7 shows the results for our anomaly detection engine when using the same GME open price data in the `GME_array.csv` file.

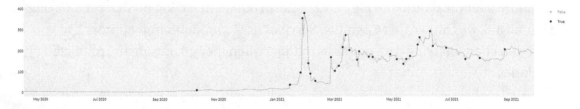

Figure 18-7. *Our anomaly detection engine focuses on change points rather than values outside an expected model, and so it picks up a different set of results than Cognitive Services*

We will save further analysis of these two engines for the next chapter.

Conclusion

Over the course of this chapter, we looked at anomaly detection offerings across three major cloud providers. From there, we learned how to configure the Azure Cognitive Services Anomaly Detector service and used a website to try out the Anomaly Detector service, comparing one real-life dataset to understand how different services can come up with differing definitions of what constitutes an outlier.

In the next chapter, we will dive into single-series time series outlier detection with Azure vs. our anomaly detection service, gaining a better understanding of how our service stacks up.

## CHAPTER 19

Performing a Bake-Off

In this final chapter of the book, we will compare our anomaly detector service to the Azure Cognitive Services Anomaly Detector service. Importantly, the intent of this chapter is not to prove conclusively which is better; it is instead an opportunity to see how two services stack up and see if our service is competitive with a commercially available option. We will begin by laying out the parameters of comparison. Then, we will perform the bake-off, comparing how the two services handle the same datasets. Finally, we will close with a few thoughts on how we can improve our anomaly detection engine.

Preparing the Comparison

Before we write code to compare the Azure Cognitive Services Anomaly Detector service to our service, we want to make as reasonable a comparison as possible. This means that we want to find datasets that are good for comparison, knowing that the Cognitive Services Anomaly Detector service does not handle non-time series data. Furthermore, when it comes to data, there is an important question to ask: Do we aim for supervised or unsupervised comparison sets?

Supervised vs. Unsupervised Learning

Supervised learning means that our training data includes not only the information we wish to feed into a model for prediction but also a *label* that represents the correct value. By contrast, *unsupervised learning* happens when we only have the inputs, not the correct outputs. When it comes to anomaly detection, we are typically solving an unsupervised learning problem: a human typically does not have the combination of time, inclination, or ability to classify every data point in a series as anomalous or non-anomalous. Therefore, we usually do not have enough information to confirm whether the outliers our engine comes up with are actual anomalies or just noise—that

© Kevin Feasel 2022
K. Feasel, *Finding Ghosts in Your Data*, https://doi.org/10.1007/978-1-4842-8870-2_19

determination is ultimately something our callers would need to make based on their skills and experiences. The downside to having a person act as arbiter of whether a particular data point is anomalous is that this will introduce bias into the judgment compared to defining a priori what the correct answers are. In other words, if we happen to have predefined labels before performing anomaly detection, we remove the bias of a judge leaning one way or the other as a result of a particular engine choosing a particular data point as either an outlier or an inlier. Even so, we should still understand that there may be some level of bias in the labeling process. That bias will not come from a desire to see one particular engine perform better than another, but it may still be there in the choices someone makes. For example, if I classify a news article as belonging either to the sports category or the finance category, my act of classification brings with it some level of bias. That bias likely would not show up if I have an article about a baseball team beating its opponent in extra innings last night, nor would it show up with an article about the merger of two retail firms. It may show up, however, in an article about a baseball team's owner negotiating a new stadium lease with the city government. When there is some level of ambiguity in selection, my personal preferences and in-built decision-making can alter the results of this classification, deciding whether it belongs in the sports category (because it has to do with a sports franchise) or the finance category (because it has to do with public revenues and financial negotiations). In short, there is no way to get around bias completely, but we can at least use labeled comparison data to minimize the risk that we grade results based on which engine we prefer rather than some other criterion. For this reason, we will emphasize datasets with labeled anomaly points. Neither engine will get to see how any data points are labeled, but we will use the labels to help us score the results.

Choosing Datasets

The Cognitive Services Anomaly Detector service works with time series data, so we will need to limit ourselves to labeled time series datasets. Fortunately for us, the Numenta Anomaly Benchmark (NAB) includes a variety of labeled single-series time series datasets we can use. You can access NAB at `https://github.com/numenta/NAB` and see the variety of datasets, as well as scores for other detection engines. The sources available include a mixture of artificial datasets as well as stock prices, web traffic data, and internal Amazon Web Services metrics. Most important for our purposes is that the shape of the data matches what we need for our anomaly detection engine as well as the one in Cognitive Services: one feature for time and one feature for value.

Scoring Results

The final consideration we need to make is how we will score the results. There are four possible outcomes for each data point we compare. The first is that the anomaly detection engine indicated that a particular data point is an outlier and it really was an anomaly. We can call these *true positives*. The second outcome is that the anomaly detection engine predicted that a particular data point is not an outlier and it really was not an anomaly, or *true negatives*. We could potentially weigh these equally, indicating that correct answers are always equally valuable. Alternatively, we could weigh correct predictions of anomalies higher than correct predictions of non-anomalies. This makes sense for two reasons. First, anomalies are, by definition, less likely to occur than non-anomalies. Second, we are more interested in anomalies than non-anomalies. For these reasons, we will weigh these two outcomes differently: a correct prediction of an anomaly rates a score of +3, and a correct prediction of a non-anomaly rates a score of 0.

The other two possible outcomes are our negative outcomes. First, we might indicate that a data point is an outlier but it turns out not to have been an anomaly. This is also known as a *false positive* and has the long-term effect of making our anomaly detection engine noisy. The downside to a noisy engine is that people are less likely to trust if the engine states that something is an outlier, as they mentally separate indications of outliers from actual anomalies. The final outcome is that we indicate a data point is not an outlier but it turns out to have been anomalous. This is also known as a *false negative* and can cause a false sense of security, where people believe everything is fine but there is actually a problem. Just as with the positive outcomes, we could weigh these negative outcomes equally, but we will not do so. We will instead weigh false negatives as a -5, as we want to discourage them. False positives, meanwhile, will get a score of -1, indicating that they are negative outcomes but not as bad as false negatives. In other words, we are more concerned with the anomaly detector engine finding all of the actual anomalies, even if it means triggering on some points that turn out not to have been anomalous.

To summarize this section:

- True positive anomaly = +3

- True negative anomaly = 0

- False positive anomaly = -1

- False negative anomaly = -5

> **Note**   The specific values we chose for scoring are somewhat arbitrary and are
> not intended to be the best or only way of scoring results. I am also choosing these
> numbers prior to reviewing any datasets or running any of the datasets we will use
> through either anomaly detection engine. The goal of this exercise is to register my
> beliefs and biases prior to performing the analysis. Your situation, including relative
> weightings of each of these outcomes, may differ considerably; even so, you can
> still use the base concept here while defining your own rubric.

Now that we know the score, let's draw some comparisons.

Performing the Bake-Off

In this section, we will create a simple Python script to call each service. This script is
in the src\comp\ folder in the source code repository and is called compare_engines.
py. Listing 19-1 shows the main() function we will use to process results for our engine
as well as the Cognitive Services Anomaly Detector. The primary goal of this script
is to read in a set of files that match a known format, process each file, and write the
file in a prescribed way into the appropriate results directory. The data_folder and
results_folder variables represent the location of datasets and where we should write
our results, respectively. If you are following along, you will want to clone the NAB Git
repo at https://github.com/numenta/NAB/ and change these to the correct values for
your setup.

Listing 19-1. The main function that drives output file generation

```python
def main():
    data_folder = "D:\\SourceCode\\NAB\\data\\"
    results_folder = "D:\\SourceCode\\NAB\\results\\"
    process_azure(data_folder, results_folder)
    process_book(data_folder, results_folder)
```

Accessing Cognitive Services via Python

Azure Cognitive Services comes with an Anomaly Detector library for Python, which you can install using the command `pip install azure-ai-anomalydetector`. You will also need to create two environment variables, one called ANOMALY_DETECTOR_KEY and the other called ANOMALY_DETECTOR_ENDPOINT. The values for these should be the same as what you used in Chapter 18 to connect to the Azure Cognitive Services Anomaly Detector demo website.

Note   Given the number of datasets and data points we will work with in this section, you might not be able to run the following code on the F0 tier of Azure Cognitive Services Anomaly Detector. When performing the analysis, I already had an S0 tier service available and made use of it, so I did not confirm whether the following will also work on the free tier.

With all of the setup complete, there are a few functions we will call to process files. The first is `process_azure()`, which we can see in Listing 19-2. This is the control method for our process. After creating an Anomaly Detector client, the function loops through each CSV file in the data directory. For each file, we want to know three things: its absolute path (`input_file`), its path relative to the base data directory (`file_location`), and the file name (`file_name`).

Listing 19-2. The function to process datasets through the Azure Cognitive Services Anomaly Detector

```
def process_azure(data_folder, results_folder):
  ANOMALY_DETECTOR_KEY = os.environ["ANOMALY_DETECTOR_KEY"]
  ANOMALY_DETECTOR_ENDPOINT = os.environ["ANOMALY_DETECTOR_ENDPOINT"]
  client = AnomalyDetectorClient(AzureKeyCredential(ANOMALY_DETECTOR_KEY),
  ANOMALY_DETECTOR_ENDPOINT)
  for input_file in glob.iglob(data_folder + '**\\*.csv', recursive=True):
    file_location = input_file.replace(data_folder, "")
    file_name = Path(input_file).name
    df = read_file_azure(input_file)
    try:
      response = detect_outliers_azure(client, df)
```

```
    df['is_anomaly'] = [v for i,v in enumerate(response.is_anomaly)]
    df['anomaly_score'] = 1.0 * df['is_anomaly']
    df['label'] = 1 * df['is_anomaly']
    df = df.drop('is_anomaly', axis=1)
    output_file = results_folder + "azure\\" + file_location.
    replace(file_name, "azure_" + file_name)
    write_file_azure(df, output_file)
    print('Completed file ' + file_name)
  except Exception as e:
    print('Skipping this file because Cognitive Services failed to return
    a result. {}'.format(file_location), 'Exception: {}'.format(e))
```

Armed with path information, we attempt to read the file into a Pandas DataFrame. Listing 19-3 shows the code for the read_file_azure() function.

Listing 19-3. Read a file into a DataFrame

```
def read_file_azure(input_file):
  input_df = pd.read_csv(input_file)
  input_df['timestamp'] = pd.to_datetime(input_df['timestamp'])
  return input_df
```

Once we have the DataFrame in place, we call detect_outliers_azure(), the function that shapes our data and calls the Anomaly Detector service. Listing 19-4 shows the contents of the detect_outliers_azure() function. We will set the sensitivity to 55 and max anomaly rate to 0.25. Then, we need to process each row in the DataFrame and turn each element into a TimeSeriesPoint() for the Anomaly Detector service, containing the timestamp in a specific format as well as the value. We will call the DetectRequest() function to make our call and return the results of detect_entire_series().

Listing 19-4. Calling the Anomaly Detector service in Python

```
def detect_outliers_azure(client, df):
  sensitivity = 55
  max_anomaly_ratio = 0.25
  series = []
  for index, row in df.iterrows():
```

```
    series.append(TimeSeriesPoint(timestamp = dt.datetime.strftime(row[0],
    '%Y-%m-%dT%H:%M:%SZ'), value = row[1]))
request = DetectRequest(series=series, custom_interval=5,
sensitivity=sensitivity, max_anomaly_ratio=max_anomaly_ratio)
return client.detect_entire_series(request)
```

Note   The Azure Cognitive Services Anomaly Detector service will not process more than 8640 points in a single series. To simplify the comparison, we will ignore files with more than 8640 data points.

Assuming the call to detect outliers succeeds, we return to `process_azure()` in Listing 19-2 and strip off the `is_anomaly` value—which will be a Boolean `True` or `False`—and attach it to our DataFrame. From there, we shape the output dataset the way we need it: a timestamp, the value, an anomaly score (float ranging from 0 to 1), a label (0 or 1, with 1 representing an outlier), and nothing else. Once we have the DataFrame the way we need it, we call `write_file_azure()`, which writes the file out to disk in the results directory. Listing 19-5 shows the code to write out a file, ensuring that all directories exist before attempting to write the file. We also ensure that we do not write out the index, as we do not need it.

Listing 19-5. Writing a file to disk

```
def write_file_azure(df, output_file):
  output_filepath = Path(output_file)
  output_filepath.parent.mkdir(parents=True, exist_ok=True)
  df.to_csv(output_file, index=False)
```

Accessing Our API via Python

The technique for processing data using our anomaly detector engine is very similar, in that we have the same set of processing functions, though they will behave a little differently. We start with Listing 19-6, showing the `process_book()` function. This function sets our server URL and method as well as picks the same sensitivity score and max fraction of anomalies we used for Azure. We also set the debug flag to `True` so that we can get the cutoff point for what constitutes an outlier.

Listing 19-6. The function to process data files using our anomaly detection engine

```python
def process_book(data_folder, results_folder):
  server_url = "http://localhost/detect"
  method = "timeseries/single"
  sensitivity_score = 55
  max_fraction_anomalies = 0.25
  debug = True
  for input_file in glob.iglob(data_folder + '**\\*.csv', recursive=True):
    file_location = input_file.replace(data_folder, "")
    file_name = Path(input_file).name
    input_data = read_file_book(input_file)
    df = detect_outliers_book(server_url, method, sensitivity_score, max_
    fraction_anomalies, debug, input_data)
    output_file = results_folder + "book\\" + file_location.replace(file_
    name, "book_" + file_name)
    write_file_book(df, output_file)
    print('Completed file ' + file_name)
```

For each file in the data folder, we call `read_file_book()` to return an appropriate DataFrame. Listing 19-7 includes the code for this function. The function is similar to the `read_file_azure()` function we saw previously, but it adds a key column as well as ensures that we have a dt column that is of type datetime.

Listing 19-7. Read file contents into a DataFrame.

```python
def read_file_book(input_file):
  input_df = pd.read_csv(input_file)
  input_df['key'] = input_df.index
  input_df['dt'] = pd.to_datetime(input_df['timestamp'])
  return input_df
```

After getting an input DataFrame, we pass that to the `detect_outliers_book()` function, which does most of the heavy lifting. Listing 19-8 includes the code for this function.

Listing 19-8. Call the anomaly detection engine for single-series time series data and shape the results for printing.

```
def detect_outliers_book(server_url, method, sensitivity_score, max_
fraction_anomalies, debug, input_df):
  input_data_set = input_df[['key', 'dt', 'value']].to_
  json(orient='records', date_format='iso')
  full_server_url = f"{server_url}/{method}?sensitivity_score={sensitivity_
  score}&max_fraction_anomalies={max_fraction_anomalies}&debug={debug}"
  r = requests.post(
    full_server_url,
    data=input_data_set,
    headers={"Content-Type": "application/json"}
  )
  res = json.loads(r.content)
  cutoff = res['debug_details']['Outlier determination']
['Sensitivity score']
  df = pd.DataFrame(res['anomalies'])
  df = df.drop('key', axis=1)
  df['anomaly_score'] = 0.5 * df['anomaly_score'] / cutoff
  # If anomaly score is greater than 1, set it to 1.0.
  df.loc[df['anomaly_score'] > 1.0, 'anomaly_score'] = 1.0
  df['label'] = 1 * df['is_anomaly']
  df = df.drop('dt', axis=1)
  return df
```

This function first uses the requests library to call our anomaly detection API. We load the JSON results into res and then pull out the sensitivity score to see where our cutoff point is for determining whether a given data point is an outlier. With this information, we build a rough calculation of anomaly score, normalizing our cutoff point to 0.5 and ranging up to 1.0 if the actual value is at least twice the cutoff score. Finally, we drop the columns we do not need and return our output DataFrame. Returning to the process_book() function, we take the output DataFrame and call write_path_book(), sending the results to disk as we see in Listing 19-9.

Listing 19-9. Writing each scored dataset to disk in the proper directory

```
def write_file_book(df, output_file):
    output_filepath = Path(output_file)
    output_filepath.parent.mkdir(parents=True, exist_ok=True)
    df.to_csv(output_file, index=False)
```

Now that we have all of the files for both anomaly detection engines in place, we can generate comparisons.

Dataset Comparisons

To create comparisons, we will need to do three things. The first thing is to modify the config\thresholds.json file in the NAB repository. This is a JSON file containing information about each test. Listing 19-10 provides the JSON necessary to enable NAB to analyze the two anomaly detection engines. It is not necessary to put these in alphabetical order, just that they show up in the file. It is also important to keep the threshold values at 0.5, as NAB needs to know the cutoff point between non-outliers and outliers.

Listing 19-10. Add these JSON blocks to your thresholds.json file.

```
"azure": {
  "reward_low_FN_rate": {
    "score": -146.0,
    "threshold": 0.5
  },
  "reward_low_FP_rate": {
    "score": -73.0,
    "threshold": 0.5
  },
  "standard": {
    "score": -73.0,
    "threshold": 0.5
  }
},
"book": {
```

```
    "reward_low_FN_rate": {
      "score": -232.0,
      "threshold": 0.5
    },
    "reward_low_FP_rate": {
      "score": -116.0,
      "threshold": 0.5
    },
    "standard": {
      "score": -116.0,
      "threshold": 0.5
    }
  },
```

The second thing to do is to change the `requirements.txt` file in the NAB project. This project was released in 2015, meaning that the requirements are for versions of packages that are seven or more years old. Remove all of the specific versions of packages, leaving just the package names (like `Cython`, `pandas`, and `simplejson`), and the parts of NAB we want to use should still work on current versions of Python.

After that, Listing 19-11 shows how to run each scoring command. These are command-line operations and should be performed in the root directory of the NAB repository.

Listing 19-11. Score and normalize results for each anomaly detection engine. Run this in the NAB repository rather than the finding-ghosts-in-your-data repository

```
python run.py -d book --score --normalize
python run.py -d azure --score --normalize
```

This will review each result file, compare the detection engine against actual market outliers, and generate a set of outputs for each file, including counts of true positives, true negatives, false positives, and false negatives. In case you do not want to run everything yourself, the `comp\NAB\results` folder contains the scoring details for both engines. Listing 19-12 shows how to load this data into Python for further analysis.

Listing 19-12. Summarize the results for two anomaly detection services against the NAB datasets. Run this in the comp\NAB\results\ directory of the finding-ghosts-in-your-data repository

```
import pandas as pd
book_df = pd.read_csv("book_standard_scores.csv")
azure_df = pd.read_csv("azure_standard_scores.csv")
book_totals = book_df[book_df['Detector'] == "Totals"]
azure_totals = azure_df[azure_df['Detector'] == "Totals"]
```

Figure 19-1 shows these counts for our anomaly detection engine vs. the Azure Cognitive Services Anomaly Detector engine.

```
>>> book_totals
    Detector Profile File  Threshold        Score    TP      TN     FP     FN  Total_Count
58    Totals    NaN  NaN        NaN  -3127.981021  3914  269435  29912  29581       332842
>>> azure_totals
    Detector Profile File  Threshold        Score    TP      TN     FP     FN  Total_Count
45    Totals    NaN  NaN        NaN   -721.299528  2322  112364   7490  10720       132896
>>> █
```

Figure 19-1. *Reviewing the results for two detection engines*

We can see that our engine handled approximately 2.5 times as many data points as the Azure engine. The reason for this is that we ignored any files with more than 8640 data points when making requests of the Cognitive Services engine. Using our formula of 3 points for a true positive, -1 for a false positive, and -5 for a false negative, our engine scores an average of -0.499 per data point. Cognitive Services scores a total of -0.407 per data point, making Cognitive Services the better choice here. Limiting ourselves to just the files that Cognitive Services used, the score improves a little to -0.484 per data point but is still worse than the alternative.

Lessons Learned

The opportunity to compare our anomaly detection engine vs. a paid product is quite useful. We were able to perform comparisons against a combination of artificial and real time series datasets, knowing that our engine had never seen this data before and therefore we could not cheat toward a nicer-looking solution. Returning back to our single-series time series solution, we focused on change point detection for outlier detection. This provides us some speculative insight for why our engine did not perform as well in this test, as we intentionally designed for noticing changes in trends rather than single anomalous data points. Taking our current engine back to the drawing board, we could potentially perform significantly better by introducing other techniques, such as ARIMA-style modeling techniques or SAX, to our single-series time series detector.

Another thing to keep in mind is that even though we performed worse than Cognitive Services, the difference was not enormous. This is a positive sign, as it indicates that there can be value in improving our anomaly detection engine, which is the focus of the next section.

Making a Better Anomaly Detector

Throughout the course of this book, we took a purposeful approach to building an anomaly detection engine. The single biggest imperative we followed throughout was to ensure that users did not need to have any systemic knowledge of the distribution of their data, including whether the data follows a particular statistical distribution, how many (if any) outliers they might have, and what are the correct cutoff points for various statistical functions given their input data. This is so core to the approach that I would not even think of it as a restriction; it is instead a guiding principle of the engine we have built. That said, there are restrictions we could loosen and techniques we could implement in order to improve the overall product quality.

Increasing Robustness

During development, we introduced some restrictions whose design was to simplify the code and make it easier for people to understand. This simplification comes at the cost of engine robustness. For example, when working with multi-series time series data, we made the assumption that all series had exactly the same timestamps and collection intervals. This is not a terrible assumption to make, but it does reduce the robustness of our solution and introduce problems if one series has different collection intervals. A solution here could be to build a process that synthesizes a common collection interval across each series. The idea would be to imagine each data point occurring on a timeline and linearly interpolating between data points in each series. Then, we would cut the timeline at the same spot for each series and find the synthetic value at each cut.

Another method to increase system robustness would be to handle bad data, such as cases when a single data point has an invalid value. At present, the detection engine requires that all data points' values be valid floats. Sometimes, however, bad data can sneak in: NULL or NA representing missing data, characters like X representing invalid samples, or other mistakes that make it into real-world datasets. The exact technique you would want to use to handle bad data would depend on your company's circumstances and use case. You might drop invalid rows, interpolate results (especially for time series data), remove non-numeric characters, or otherwise transform the data before running any outlier detection tests.

Extending the Ensembles

The next way to improve our anomaly detection engine would be to extend each ensemble. In this chapter, we looked solely at single-series time series analysis, the one mechanism for which we did not create an explicit ensemble. Ensembles are a great technique for making reasonable guesses under uncertainty, particularly when the ensembles operate from different assumptions and perform different statistical operations. As we learned in Chapter 6, well-chosen ensembles allow us to maximize the benefits from multiple moderately effective tests. Furthermore, for each class of outlier detection technique we implemented, there are a large number of statistical techniques and approaches we did not include in our ensemble. Reviewing some of these other techniques and incorporating them—whether by mere addition or by swapping out some other technique—could lead to better results overall.

Training Parameter Values

A final way to improve our anomaly detection engine would be to make better-informed choices regarding individual parameter values, such as cutoff points for techniques like COF, LOCI, and SAX. The difficulty in making better-informed choices is that doing so requires one of two things: either your users understand their data well enough to provide relevant inputs or you know your users' data well enough that you can define relevant inputs. The former clashes with the ethos of our project, but the latter is viable if quite difficult. The reason for the difficulty is based in our assumption that this is a general-purpose anomaly detection service, meaning we can get data from any source system and treat it the best we can. If you have a good idea of what datasets your users will submit, however, you could take advantage of this knowledge. Suppose, for example, that you have in your possession labeled company data that indicates anomalies, similar to the NAB datasets we worked with in this chapter. With this labeled data, you could treat the various parameters we've hard-coded throughout the book as *hyperparameters*, that is, parameters whose values you set before training a model. You would also need a scoring function, which could be the one we used in this chapter. Armed with both of these, you could find the set of cutoff points that work best for the datasets you trained your engine against. The risk here is that you assume future datasets will follow from old datasets in terms of optimal cutoff points, but to the extent that this holds true, you could end up with significantly better results than a general-purpose engine that needs to satisfy all possible datasets without having any knowledge of the data beforehand.

Conclusion

In this chapter, we learned how our anomaly detection engine compares to a commercially available engine in Azure Cognitive Services Anomaly Detector, running against a battery of artificial and real datasets. Although the commercial product did beat out our solution, the difference was not enormous. Further, we gained some insight into how to improve the engine. We ended this chapter with several ideas around how to improve the engine quality as well as the user experience with our service.

APPENDIX

Bibliography

The following resources were helpful in shaping the course of this book. Although not all of these works received explicit citations, they all provided useful guidance and background information which I aimed to distill.

Aggarwal, Charu C. *Outlier Analysis*. 2nd ed., Springer, 2017.

Aminikhanghahi, Samaneh and Diane J. Cook. "A Survey of Methods for Time Series Change Point Detection." *Knowl Inf Syst*. 51(2), 2017 May, pp. 339–367, doi 10.1007/s10115-016-0987-z.

Brownlee, Jason. "A Gentle Introduction to Normality Tests in Python." *Machine Learning Mastery*, May 2018. machinelearningmastery.com/a-gentle-introduction-to-normality-tests-in-python.

Brownlee, Jason. "A Gentle Introduction to Statistical Data Distributions." *Machine Learning Mastery*, June 2018. machinelearningmastery.com/statistical-data-distributions.

Brownlee, Jason. "How to Transform Data to Better Fit the Normal Distribution." *Machine Learning Mastery*, May 2018. machinelearningmastery.com/how-to-transform-data-to-fit-the-normal-distribution.

Buczyński, Sebastian. *Implementing the Clean Architecture*. n.p.: Author, 2020.

Chiang, Alvin, Esther David, Yuh-Jye Lee, Guy Leshem, and Yi-Re Yeh. "A study on anomaly detection ensembles." *Journal of Applied Logic*, 21, May 2017, pp. 1–13, doi 10.1016/j.jal.2016.12.002.

Damodaran, Aswath. "A Review of Statistical Distributions." pages.stern.nyu.edu/~adamodar/pc/blog/probdist.pdf.

Durante, Fabrizio and Carlo Sempi. "Copula Theory: An Introduction." In Jaworski, P., Durante, F., Härdle, W., Rychlik, Td. (eds). *Copula Theory and Its Applications. Lecture Notes in Statistics*, vol 198, 2010, doi 10.1007/978-3-642-12465-5_1.

Erten, Aischa. *Gestalt in Photography*. n.p.: Author, 2021.

© Kevin Feasel 2022
K. Feasel, *Finding Ghosts in Your Data*, https://doi.org/10.1007/978-1-4842-8870-2

Fernandez, Angela, Juan Bella, and Jose R. Dorronsoro. "Supervised outlier detection for classification and regression." *Neurocomputing*, vol. 486, May 2022, pp. 77–92, doi 10.1016/j.neucom.2022.02.047.

Grubbs, Frank E. "Procedures for Detecting Outlying Observations in Samples." *Technometrics*, vol. 11, no. 1, 1969, pp. 1–21. JSTOR, doi 10.2307/1266761.

Guha, Sudipto, Nina Mishra, Gourav Roy, and Okke Schrijvers. "Robust Random Cut Forest Based Anomaly Detection on Streams." Proceedings of the 33rd International Conference on Machine Learning, PMLR, 48, 2016, pp. 2712–2721, doi 10.5555/3045390.3045676.

Hall, Robert C. "Properties of Data Sets that Conform to Benford's Law." 2019. vixra. org/abs/1906.0131.

Hampel, Frank. "Robust Statistics: A Brief Introduction and Overview." 2001. `www.researchgate.net/publication/252706505_Robust_statistics_A_brief_introduction_and_overview`.

Hayek, Friedrich A. *The Sensory Order*. Martino Publishing, 2014.

Haynes, Kaylea, Idris A. Eckley, and Paul Fearnhead. "Efficient penalty search for multiple changepoint problems." *arXiv: Computation*, 2014, doi 10.48550/arXiv.1412.3617.

Kenton, Will. "Copula." *Investopedia*, 2021. `www.investopedia.com/terms/c/copula.asp`.

Keogh, Eamonn, Jessica Lin, and Ada Fu. "HOT SAX: Finding the Most Unusual Time Series Subsequence: Algorithms and Applications." *Proceeding of the 5th IEEE International Conference on Data Mining (ICDM)*, 2005 Nov, pp. 226–233, doi 10.1109/ICDM.2005.79.

Koffka, Kurt. *Principles of Gestalt Psychology*. Mimesis International, 2014.

Li, Zheng, Yue Zhao, Nicola Botta, Cezar Ionescu, and Xiyang Hu. "COPOD: Copula-Based Outlier Detection." *IEEE International Conference on Data Mining*, March 2021.

Lin, Jessica, Eamonn Keogh, Stefano Lonardi, and Pranav Patel. "Finding Motifs in Time Series." *KDD Proceedings of the Second Workshop on Temporal Data Mining*, 2002.

Liu, Fei Tony, Kai Ming Ting, and Zhi-Hua Zhou. "Isolation Forest." *2008 Eighth IEEE International Conference on Data Mining*, 2008 pp. 413–422, doi 10.1109/ICDM.2008.17.

McGonagle, John, Geoff Pilling, and Andrei Dobre. "Gaussian Mixture Model." Brilliant.org. brilliant.org/wiki/gaussian-mixture-model.

Mair, Patrick and Rand Wilcox. "Robust Statistical Methods in R using the WRS2 Package". *Behav Res* 52, 2020, pp. 464-488. doi 10.3758/s13428-019-01246-w.

Mehrotra, Kishan G., Chilukuri K. Mohan, and HuaMing Huang. *Anomaly Detection Principles and Algorithms*. Springer, 2017.

Mohan, Chilukuri K. and Kishan Mehrotra. "Anomaly detection in banking operations." 2017.

Moitra, Ankur. *Algorithmic Aspects of Machine Learning*. n.p.: Author, 2015. ocw.mit.edu/courses/18-409-algorithmic-aspects-of-machine-learning-spring-2015/resources/mit18_409s15_bookex/

NIST/SEMATECH e-Handbook of Statistical Methods, www.itl.nist.gov/div898/handbook. doi 10.18434/M32189.

Nowak-Brzezinska, Agnieszka and Czeslaw Horyn. "Outliers in rules – the comparison of LOF, COF and KMEANS algorithms." *Procedia Computer Science*, vol. 176, 2020, pp. 1420–1429, doi 10.1016/j.procs.2020.09.152.

Pakdaman, Mahdi. "Introduction to Copula Functions, part 2." 2011. people.cs.pitt.edu/~pakdaman/tutorials/Copula02.pdf.

Papadimitriou, Spiros, Hiroyuki Kitagawa, and Phillip B. Gibbons. "LOCI: Fast Outlier Detection Using the Local Correlation Integral." *Proceedings of the 19th International Conference on Data Engineering* (Cat. No.03CH37405), 2003, pp. 315–326.

Patel, Pranav, Eamonn Keogh, Jessica Lin, and Stefano Lonardi. "Mining Motifs in Massive Time Series Databases." *2002 IEEE International Conference on Data Mining*, 2002, pp.370–377, doi 10.1109/ICDM.2002.1183925.

Raschka, Sebastian. "Dixon's Q test for outlier identification—A questionable practice." *Sebastian Raschka*, July 2014. sebastianraschka.com/Articles/2014_dixon_test.html.

Ripley, B.D. "Robust Statistics." 2005. www.stats.ox.ac.uk/~ripley/StatMethods/Robust.pdf.

Sakia, R. M. "The Box-Cox transformation technique: a review." *The Statistician* 41, 1992, pp. 169–178.

Shetty, Charanaj. "Time Series Models: AR, MA, ARMA, ARIMA." *Towards Data Science*, 2020. towardsdatascience.com/time-series-models-d9266f8ac7b0

Tang, Jiang, Zhixiang Chen, Ada Wai-chee Fu, and David W. Cheung. "Enhancing Effectiveness of Outlier Detections for Low Density Patterns." *Pacific-Asia Conf. on Knowledge Discovery and Data Mining (PAKDD)*. Taipei. 2002. pp. 535–548. doi 10.1007/3-540-47887-6_53.

Teugels, Jef L. and Giovanni Vanroelen. "Box-Cox Transformations and Heavy-Tailed Distributions." *Journal of Applied Probability*, vol. 41, 2004, pp. 213–27. JSTOR, www.jstor.org/stable/3215978.

Truong, Charles, Laurent Oudre, and Nicolas Vayatis. "Selective review of offline change point detection methods." *Signal Processing*, 167, 2020 Feb, doi 10.1016/j. sigpro.2019.107299.

Uberoi, Anannya. "K-Nearest Neighbours." Geeks for Geeks. www.geeksforgeeks. org/k-nearest-neighbours.

VanderPlas, Jake. *Python Data Science Handbook*. O'Reilly Media, Inc., 2016.

Wagemans, Johan et al. "A Century of Gestalt Psychology in Visual Perception I. Perceptual Grouping and Figure-Ground Organization." Psychological Bulletin, 138(6), 2012 Nov, pp. 1172–1217. doi 10.1037/a0029333.

Wambui, Gachomo Dorcas, Gichuhi Anthony Waititu, and Anthony Wanjoya. "The Power of Pruned Exact Linear Time (PELT) Test in Multiple Changepoint Detection." *American Journal of Theoretical and Applied Statistics*, 4(6), 2015 Nov, pp. 581–586, doi 10.11648/j.ajtas.20150406.30.

Warren, Rik, Robert E. Smith, and Anne K. Cybenko. "Use of Mahalanobis Distance for Detecting Outliers and Outlier Clusters in Markedly Non-Normal Data: A Vehicular Traffic Example." June 2011. apps.dtic.mil/sti/pdfs/ADA545834.pdf

Wertheimer, Max. "Experimentelle Studien über das Sehen von Bewegung." *Zeitschrift für Psychologie und Physiologie der Sinnesorgane*, 61, pp. 161–265.

Wiecki, Thomas. "An intuitive, visual guide to copulas." 2018. twiecki.io/ blog/2018/05/03/copulas.

Zhang, Mingda. "Time Series: Autoregressive models. AR, MA, ARMA, ARIMA." 2018, people.cs.pitt.edu/~milos/courses/cs3750/lectures/class16.pdf.

Index

A

Amazon Web Services (AWS),
313–314, 326

Anderson-Darling test, 119, 120, 122,
136, 156

Anomalies
building detector, 7
industry/field, 15
finance, 8–10
graphical example, 5
outlier, 3
medicine field, 11
vs. noise, 4, 5
outlier predictions *vs.* data point's
actual status, 7, 8
outliers, 22
sports
BABIP, 12
defining, 11
ERA, 13, 14
FIP, 13
theoretical model, 6, 7
web analytics, 14, 15

Anomaly detection, 1
clustering, 16
design
alert systems administrators/
developers, 18, 19
humans handle anomalies, 19, 20
supervised learning, 21
modeling techniques, 17
statistical, 16

Anomaly detection engine, 311, 322
extend ensemble, 338
increasing robustness, 337
statistical functions, 337
training parameter value, 338

Anomaly detection service, 18, 20

Anomaly detector service, 72, 325

ARIMA-style modeling, 337

assert() function, 87

Autocorrelation, 233, 234, 241, 277

Autoregression (AR), 241, 242

Average of Maximum (AOM), 209

Azure Cognitive Services, 314, 315, 325
account setup
anomaly detectors, 317
API keys, 320
creating services, 318, 319
endpoint URL, 319
subscription, 316
writing—Azure pricing, 316
demo applications, 320–322

Azure Cognitive Services Anomaly
Detector, 311, 313–323, 339

Azure Stream Analytics, 315

B

BaseModel, 76

Batting Average on Balls in Play (BABIP), 12

Bayesian information criterion (BIC),
172, 179

BigQuery ML, 315, 316

Box-Cox transformation, 123, 138, 266

345

K. Feasel, *Finding Ghosts in Your Data*, https://doi.org/10.1007/978-1-4842-8870-2